Welcome to *My Math* – your very own math book!
You can write in it – in fact, you are encouraged to
write, draw, circle, explain, and color as you explore
the exciting world of mathematics. Let's get started.
Grab a pencil and finish each sentence.

My name is _____.

My favorite color is _____.

My favorite hobby or sport is _____.

My favorite TV program or video game is

_____.

My favorite class is _____.

McGraw Hill Education

mhmymath.com

STEM McGraw-Hill is committed to providing
instructional materials in Science, Technology, Engineering, and
Mathematics (STEM) that give all students a solid foundation,
one that prepares them for college and careers in the 21st
century.

Send all inquiries to:
McGraw-Hill Education
8787 Orion Place
Columbus, OH 43240

ISBN: 978-0-07-905763-1 (**Volume 1**)
MHID: 0-07-905763-2

Printed in the United States of America.

 7 8 9 10 11 12 QSX 24 23 22 21 20

McGraw-Hill My Math

Grade 4 • Volume 1

Authors:

Carter • Cuevas • Day • Malloy

Altieri • Balka • Gonsalves • Grace • Krulik • Molix-Bailey

Moseley • Mowry • Myren • Price • Reynosa • Santa Cruz

Silbey • Vielhaber

McGraw Hill Education

GO digital ▶ connectED.mcgraw-hill.com

▶ Log In

1 Go to
connectED.mcgraw-hill.com.

2 Log in using your username
and password.

3 Click on the
Student
Edition icon
to open the
Student
Center.

Grade 4

▶ Go to the Student Center

4 Click on Menu, then click on the **Resources** tab
to see all of your online resources arranged by
chapter and lesson.

5 Click on the **eToolkit** in the Lesson Resources
section to open a library of eTools and virtual
manipulatives.

6 Look here to find any assignments or
messages from your teacher.

7 Click on the **eBook** to open your online
Student Edition.

▶ Explore the eBook!

8 Click the **speaker icon** at the top of the eBook pages to hear the page read aloud to you.

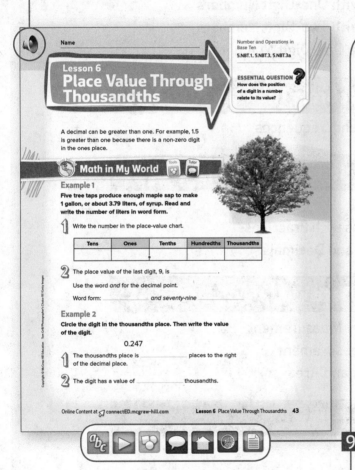

More resources can be found by clicking the icons at the bottom of the eBook pages.

 Practice and review your Vocabulary.

 Animations and videos allow you to explore mathematical topics.

 Explore concepts with eTools and virtual manipulatives.

 Personal Tutors are online virtual teachers that walk you through the steps of the lesson problems.

 eHelp helps you complete your homework.

 Explore these fun digital activities to practice what you learned in the classroom.

 Worksheets are PDFs for Math at Home, Problem of the Day, and Fluency Practice.

Contents in Brief
Organized by Domain

Number and Operations in Base Ten

Chapter 1   Place Value
Chapter 2   Add and Subtract Whole Numbers
Chapter 3   Understand Multiplication and Division
Chapter 4   Multiply with One-Digit Numbers
Chapter 5   Multiply with Two-Digit Numbers
Chapter 6   Divide by a One-Digit Number

Operations and Algebraic Thinking

Chapter 7   Patterns and Sequences

Number and Operations — Fractions

Chapter 8   Fractions
Chapter 9   Operations with Fractions
Chapter 10 Fractions and Decimals

Measurement and Data

Chapter 11 Customary Measurement
Chapter 12 Metric Measurement
Chapter 13 Perimeter and Area

Geometry

Chapter 14 Geometry

Processes &Practices → Woven Throughout

connectED.mcgraw-hill.com

Chapter

Place Value

Number and Operations in Base Ten

ESSENTIAL QUESTION
How does place value help represent the value of numbers?

Getting Started

My Chapter Project 2
Am I Ready? 3
My Math Words 4
My Vocabulary Cards 5
My Foldable **FOLDABLES** 9

Lessons and Homework

Lesson 1 Place Value 11
Lesson 2 Read and Write Multi-Digit Numbers 17
Lesson 3 Compare Numbers 23
Lesson 4 Order Numbers 29
Check My Progress 35
Lesson 5 Use Place Value to Round 37
Lesson 6 Problem-Solving Investigation:
Use the Four-Step Plan 43

Wrap Up

My Chapter Review 49
Reflect 52
Performance Task 52PT1

Look for this!
Click online and you can watch videos that will help you learn the lessons.

Watch

Number and Operations in Base Ten

ESSENTIAL QUESTION
What strategies can I use to add or subtract?

Chapter

2 Add and Subtract Whole Numbers

Getting Started

My Chapter Project 54
Am I Ready? 55
My Math Words 56
My Vocabulary Cards 57
My Foldable **FOLDABLES** 59

Lessons and Homework

Lesson 1 Addition Properties and Subtraction Rules 61

Lesson 2 Addition and Subtraction Patterns 67

Lesson 3 Add and Subtract Mentally 73

Lesson 4 Estimate Sums and Differences 79

Check My Progress 85

Lesson 5 Add Whole Numbers 87

Lesson 6 Subtract Whole Numbers 93

Lesson 7 Subtract Across Zeros 99

Check My Progress 105

Lesson 8 Problem-Solving Investigation: Draw a Diagram 107

Lesson 9 Solve Multi-Step Word Problems 113

Wrap Up

Fluency Practice 119
My Chapter Review 121
Reflect 124
Performance Task 124PT1

connectED.mcgraw-hill.com

Understand Multiplication and Division

Number and Operations in Base Ten

ESSENTIAL QUESTION
How are multiplication and division related?

Getting Started

My Chapter Project **126**

Am I Ready? **127**

My Math Words **128**

My Vocabulary Cards **129**

My Foldable **FOLDABLES** **133**

Lessons and Homework

Lesson 1 Relate Multiplication and Division **135**

Lesson 2 Relate Division and Subtraction **141**

Lesson 3 Multiplication as Comparison **147**

Lesson 4 Compare to Solve Problems **153**

Check My Progress **159**

Lesson 5 Multiplication Properties and
Division Rules **161**

Lesson 6 The Associative Property of
Multiplication **167**

Lesson 7 Factors and Multiples **173**

Lesson 8 Problem-Solving Investigation:
Reasonable Answers **179**

Wrap Up

My Chapter Review **185**

Reflect **188**

Performance Task **188PT1**

eHelp **Look for this!**
Click online and you can get more help while doing your homework.

Chapter

4 Multiply with One-Digit Numbers

Getting Started

My Chapter Project **190**

Am I Ready? **191**

My Math Words **192**

My Vocabulary Cards **193**

My Foldable **FOLDABLES** **195**

Lessons and Homework

Lesson 1 Multiples of 10, 100, and 1,000 .... **197**

Lesson 2 Round to Estimate Products **203**

Lesson 3 Hands On: Use Place Value to Multiply **209**

Lesson 4 Hands On: Use Models to Multiply **215**

Check My Progress **221**

Lesson 5 Multiply by a Two-Digit Number **223**

Lesson 6 Hands On: Model Regrouping **229**

Lesson 7 The Distributive Property **235**

Lesson 8 Multiply with Regrouping **241**

Lesson 9 Multiply by a Multi-Digit Number **247**

Check My Progress **253**

Lesson 10 Problem-Solving Investigation:
Estimate or Exact Answer **255**

Lesson 11 Multiply Across Zeros **261**

Wrap Up

My Chapter Review **267**

Reflect **270**

Performance Task **270PT1**

Chapter 5 Multiply with Two-Digit Numbers

Number and Operations in Base Ten

ESSENTIAL QUESTION
How can I multiply by a two-digit number?

Getting Started

My Chapter Project 272

Am I Ready? 273

My Math Words 274

My Vocabulary Cards 275

My Foldable FOLDABLES 277

Lessons and Homework

Lesson 1 Multiply by Tens 279

Lesson 2 Estimate Products 285

Check My Progress 291

Lesson 3 Hands On: Use the Distributive Property to Multiply 293

Lesson 4 Multiply by a Two-Digit Number 299

Lesson 5 Solve Multi-Step Word Problems 305

Lesson 6 Problem-Solving Investigation: Make a Table 311

Wrap Up

My Chapter Review 317

Reflect 320

Performance Task 320PT1

Look for this!
Click online and you can find tools that will help you explore concepts.

connectED.mcgraw-hill.com

Number and Operations in Base Ten

ESSENTIAL QUESTION
How does division affect numbers?

Chapter

6

Divide by a One-Digit Number

Getting Started

My Chapter Project 322

Am I Ready? 323

My Math Words 324

My Vocabulary Cards 325

My Foldable **FOLDABLES** 327

Lessons and Homework

Lesson 1 Divide Multiples of 10, 100, and 1,000 329

Lesson 2 Estimate Quotients 335

Lesson 3 Hands On: Use Place Value to Divide 341

Lesson 4 Problem-Solving Investigation:
Make a Model 347

Lesson 5 Divide with Remainders 353

Lesson 6 Interpret Remainders 359

Check My Progress 365

Lesson 7 Place the First Digit 367

Lesson 8 Hands On: Distributive Property and
Partial Quotients 373

Lesson 9 Divide Greater Numbers 379

Check My Progress 385

Lesson 10 Quotients with Zeros 387

Lesson 11 Solve Multi-Step Word Problems 393

Wrap Up

Fluency Practice 399

My Chapter Review 401

Reflect 404

Performance Task 404PT1

connectED.mcgraw-hill.com

Chapter

Patterns and Sequences

Getting Started

My Chapter Project **406**

Am I Ready? **407**

My Math Words **408**

My Vocabulary Cards **409**

My Foldable **FOLDABLES** **411**

Lessons and Homework

Lesson 1 Nonnumeric Patterns **413**

Lesson 2 Numeric Patterns **419**

Lesson 3 Sequences **425**

Lesson 4 Problem-Solving Investigation:
Look for a Pattern **431**

Check My Progress **437**

Lesson 5 Addition and Subtraction Rules **439**

Lesson 6 Multiplication and Division Rules **445**

Lesson 7 Order of Operations **451**

Check My Progress **457**

Lesson 8 Hands On: Equations with
Two Operations **459**

Lesson 9 Equations with Multiple Operations **465**

Wrap Up

My Chapter Review **471**

Reflect **474**

Performance Task **474PT1**

Tutor **Look for this!**
Click online and you
can watch a teacher
solving problems.

connectED.mcgraw-hill.com

Number and Operations - Fractions

ESSENTIAL QUESTION
How can different fractions name the same amount?

Chapter

8 Fractions

Getting Started

My Chapter Project **476**

Am I Ready? **477**

My Math Words **478**

My Vocabulary Cards **479**

My Foldable FOLDABLES **483**

Lessons and Homework

Lesson 1 Factors and Multiples **485**

Lesson 2 Prime and Composite Numbers **491**

Check My Progress **497**

Lesson 3 Hands On: Model Equivalent Fractions **499**

Lesson 4 Equivalent Fractions **505**

Lesson 5 Simplest Form **511**

Lesson 6 Compare and Order Fractions **517**

Lesson 7 Use Benchmark Fractions to
Compare and Order **523**

Check My Progress **529**

Lesson 8 Problem-Solving Investigation:
Use Logical Reasoning **531**

Lesson 9 Mixed Numbers **537**

Lesson 10 Mixed Numbers and
Improper Fractions **543**

Wrap Up

My Chapter Review **549**

Reflect **552**

Performance Task **552PT1**

connectED.mcgraw-hill.com

Chapter

9 Operations with Fractions

Number and Operations – Fractions

ESSENTIAL QUESTION
How can I use operations to model real-world fractions?

Getting Started

My Chapter Project 554

Am I Ready? 555

My Math Words 556

My Vocabulary Cards 557

My Foldable **FOLDABLES** 559

Lessons and Homework

Lesson 1 Hands On: Use Models to Add
Like Fractions 561

Lesson 2 Add Like Fractions 567

Lesson 3 Hands On: Use Models to Subtract
Like Fractions 573

Lesson 4 Subtract Like Fractions 579

Check My Progress 585

Lesson 5 Problem-Solving Investigation:
Work Backward 587

Lesson 6 Add Mixed Numbers 593

Lesson 7 Subtract Mixed Numbers 599

Check My Progress 605

Lesson 8 Hands On: Model Fractions
and Multiplication 607

Lesson 9 Multiply Fractions by Whole Numbers 613

Wrap Up

My Chapter Review 619

Reflect 622

Performance Task 622PT1

Vocab
abc

Look for this!
Click online and you can find activities to help build your vocabulary.

Chapter
10 Fractions and Decimals

Getting Started

My Chapter Project **624**

Am I Ready? **625**

My Math Words **626**

My Vocabulary Cards **627**

My Foldable **FOLDABLES** **629**

Lessons and Homework

Lesson 1 Hands On: Place Value Through Tenths
and Hundredths **631**

Lesson 2 Tenths **637**

Lesson 3 Hundredths **643**

Check My Progress **649**

Lesson 4 Hands On: Model Decimals
and Fractions **651**

Lesson 5 Decimals and Fractions **657**

Lesson 6 Use Place Value and Models to Add **663**

Lesson 7 Compare and Order Decimals **669**

Lesson 8 Problem-Solving Investigation:
Extra or Missing Information **675**

Wrap Up

My Chapter Review **681**

Reflect **684**

Performance Task **684PT1**

connectED.mcgraw-hill.com

Chapter 11 Customary Measurement

Getting Started

My Chapter Project **686**
Am I Ready? **687**
My Math Words **688**
My Vocabulary Cards **689**
My Foldable **FOLDABLES** **695**

Lessons and Homework

Lesson 1 Customary Units of Length **697**

Lesson 2 Convert Customary Units of Length **703**

Lesson 3 Customary Units of Capacity **709**

Lesson 4 Convert Customary Units of Capacity **715**

Check My Progress **721**

Lesson 5 Customary Units of Weight **723**

Lesson 6 Convert Customary Units of Weight **729**

Lesson 7 Convert Units of Time **735**

Check My Progress **741**

Lesson 8 Display Measurement Data in a
Line Plot **743**

Lesson 9 Solve Measurement Problems **749**

Lesson 10 Problem-Solving Investigation:
Guess, Check, and Revise **755**

Wrap Up

My Chapter Review **761**
Reflect **764**
Performance Task **764PT1**

connectED.mcgraw-hill.com

Measurement and Data

ESSENTIAL QUESTION
How can conversion of measurements help me solve real-world problems?

Chapter

12 Metric Measurement

Getting Started

My Chapter Project **766**

Am I Ready? **767**

My Math Words **768**

My Vocabulary Cards **769**

My Foldable FOLDABLES **773**

Lessons and Homework

Lesson 1 Metric Units of Length **775**

Lesson 2 Metric Units of Capacity **781**

Lesson 3 Metric Units of Mass **787**

Check My Progress **793**

Lesson 4 Problem-Solving Investigation:
Make an Organized List **795**

Lesson 5 Convert Metric Units **801**

Lesson 6 Solve Measurement Problems **807**

Wrap Up

My Chapter Review **813**

Reflect **816**

Performance Task **816PT1**

Measurement and Data

ESSENTIAL QUESTION
Why is it important to measure perimeter and area?

Chapter 13 Perimeter and Area

Getting Started

My Chapter Project 818

Am I Ready? 819

My Math Words 820

My Vocabulary Cards 821

My Foldable FOLDABLES 823

Lessons and Homework

Lesson 1 Measure Perimeter 825

Lesson 2 Problem-Solving Investigation:
Solve a Simpler Problem 831

Check My Progress 837

Lesson 3 Hands On: Model Area 839

Lesson 4 Measure Area 845

Lesson 5 Relate Area and Perimeter 851

Wrap Up

My Chapter Review 857

Reflect 860

Performance Task 860PT1

ESSENTIAL QUESTION
How are different ideas about geometry connected?

Chapter

14 Geometry

Getting Started

My Chapter Project 862

Am I Ready? 863

My Math Words 864

My Vocabulary Cards 865

My Foldable **FOLDABLES** 871

Lessons and Homework

Lesson 1   Draw Points, Lines, and Rays 873

Lesson 2   Draw Parallel and Perpendicular Lines 879

Check My Progress 885

Lesson 3   Hands On: Model Angles 887

Lesson 4   Classify Angles 893

Lesson 5   Measure Angles 899

Lesson 6   Draw Angles 905

Lesson 7   Solve Problems with Angles 911

Check My Progress 917

Lesson 8   Triangles 919

Lesson 9   Quadrilaterals 925

Lesson 10 Draw Lines of Symmetry 931

Lesson 11 Problem-Solving Investigation:
Make a Model 937

Wrap Up

My Chapter Review 943

Reflect 946

Performance Task 946PT1

Chapter

1 Place Value

ESSENTIAL QUESTION

How does place value help represent the value of numbers?

Our Great Outdoors

Watch

Watch a video!

Brain Builders

MY Chapter Project

What's That Cost?

Day 1

1. Work together with your group to brainstorm a list of items you normally buy.

2. Use the Internet, magazines, and newspapers to research the price of each item. In the space below, list the items in order from least expensive to most expensive.

3. Think of a question you could ask your classmates. Write your question on the lines below. *Sample questions: If you bought 2 of the most expensive item, how much would you pay? How much less does the least expensive item cost than the second most expensive item?*

Question: _____

_____

_____

4. As a group, create a poster using pictures and prices of the items. Order the items on your poster from least expensive to most expensive. Use markers to decorate your poster.

Name
....

Am I Ready?

Compare. Use >, <, or =.

1. 8,000 ◯ 8,100   **2.** 3,404 ◯ 3,044   **3.** 7,635 ◯ 7,635

Round each number to the nearest ten.

4. 24   **5.** 16   **6.** 37

_____   _____   _____

Round each number to the nearest hundred.

7. 215   **8.** 189   **9.** 371

_____   _____   _____

10. A recipe calls for 11 eggs. Write this number in word form.

_____

Write each set of numbers in order from _least_ to _greatest_.

11. 124, 139, 129   **12.** 257, 184, 321

_____   _____

13. There are twenty-five students in Cooper's class. Write this number in standard form.

_____

Shade the boxes to show the problems you answered correctly.

 How Did I Do?

1	2	3	4	5	6	7	8	9	10	11	12	13

Name ....

MY Math Words 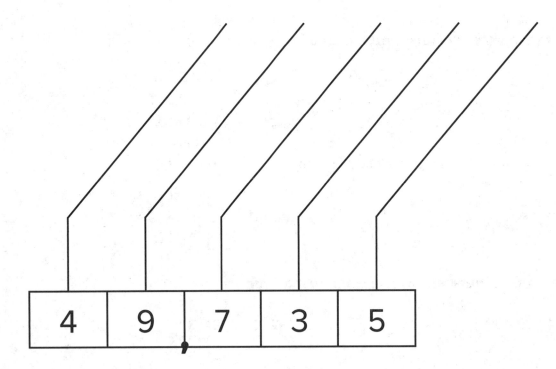 Vocab abc

Review Vocabulary

hundreds   ones   ten thousands   tens   thousands

Making Connections

Use the review vocabulary words to describe each digit in the diagram.
Then answer the question.

| 4 | 9 | 7 | 3 | 5 |

Suppose 49,735 people have tickets for a basketball game. Arena A
seats 40,000 people. Arena B seats 50,000. Which arena should host the
game? Explain how you would decide.

_____

_____

MY Vocabulary Cards

Vocab

Processes & Practices

Lesson 1-1

digit

2,340,581

2, 3, 4, 0, 5, 8, 1

Lesson 1-2

expanded form

105,073 = 100,000 + 5,000

+ 70 + 3

Lesson 1-3

is equal to (=)

1,500,000 = 1,500,000

Lesson 1-3

is greater than (>)

1,900,000 > 1,700,000

Lesson 1-3

is less than (<)

1,200,000 < 1,600,000

Lesson 1-3

number line

Lesson 1-2

period

Millions Period			Thousands Period			Ones Period		
Millions			Thousands			Ones		
hundreds	tens	ones	hundreds	tens	ones	hundreds	tens	ones
6	5	0,	0	8	4,	9	7	0

Lesson 1-1

place value

Millions			Thousands			Ones		
hundreds	tens	ones	hundreds	tens	ones	hundreds	tens	ones
6	5	0,	0	8	4,	9	7	0

Ideas for Use

- Develop categories for the words. Sort them by category. Ask another student to guess each category.

- Write a tally mark on each card every time you read or write the word. Try to use at least 10 tally marks for each word card.

The representation of a number as a *sum* that shows the value of each digit.

Write the suffix in *expanded*. Use a dictionary if you need help.

_____

Any symbol used to write a whole number.

Write a number in which the digit 6 is in the ones place and the digit 0 is in the ten thousands place.

_____

An inequality relationship showing the number on the left is greater than the number on the right.

List three other math words you can use to compare that end in -*er*.

_____

_____

Having the same value as.

The suffix -*ity* means "state or condition." What does the word *equality* mean?

_____

_____

A line with numbers on it in order at regular intervals.

Write a word problem to compare numbers.

_____

_____

_____

An inequality relationship showing the number on the left is less than the number on the right.

Write a number sentence about two sets of items in the room. Use the symbol < in your sentence.

_____

_____

The value given to a digit by its place in a number.

Write a five-digit number. Then write the place value for each digit.

_____

_____

The name given to each group of three digits in a place-value chart.

What are the names of the three periods on this card?

_____

MY Vocabulary Cards

standard form

$$3,000 + 400 + 90 + 1 = \underbrace{3,491}_{\text{standard form}}$$

word form

16,499 = sixteen thousand, four hundred ninety-nine

Ideas for Use

- Draw or write examples for each card. Be sure your examples are different from what is shown on each card.

- Write the name of each lesson on the front of each blank card. Write a few study tips for each lesson on the back of each card.

The form of a number that uses written words.

Write a five-digit number using numerals. Then rewrite it in word form.

_____

_____

_____

The usual way of writing a number that shows only its *digits*, no words.

One meaning of *standard* is "normal or usual". How can knowing this help you remember the definition for *standard form*?

_____

_____

MY Foldable

ones ×1

tens ×10

hundreds ×10 ×10

thousands ×10 ×10 ×10

ten thousands ×10 ×10 ×10 ×10

hundred thousands ×10 ×10 ×10 ×10 ×10

millions ×10 ×10 ×10 ×10 ×10 ×10

9

8

7

6

5

4

3

2

1

1

2

3

Lesson 1
Place Value

ESSENTIAL QUESTION

How does place value help represent the value of numbers?

A **digit** is any symbol used to write a whole number. The value given to a digit by its place in a number is called **place value**. A place-value chart shows the value of the digits in a number.

Math in My World

Tools Watch Tutor

Example 1

The average pencil can draw a line that is almost 184,800 feet long. What is the value of the highlighted digit?

The place-value chart shows 184,800.

Thousands Period			Ones Period		
hundreds	tens	ones	hundreds	tens	ones
1	8	4	8	0	0

 100,000
1 × 100,000

80,000
8 × 10,000

 4,000
4 × 1,000

800
8 × 100

 0
0 × 10

 0
0 × 1

The highlighted digit, 8, is in the _____ place.

So, its value is _____ .

A digit in each place represents ten times what it would represent in the place to its right.
When 8 is in the ten thousands place it has a value of 80,000.
If 8 were in the hundred thousands place, it would have a value of 10 × 80,000, or 800,000.

Example 2

There are 2,419,200 seconds in four weeks.
How does the value of the digit in the hundreds
place change if this digit were to move to each
of the four places to its left?

Millions Period			Thousands Period			Ones Period		
hundreds	tens	ones	hundreds	tens	ones	hundreds	tens	ones
		2	4	1	9	2	0	0

The digit in the hundreds place is _____.

It has a value of _____.

If this digit were to move to the thousands place, it would have
a value of _____.

If this digit were to move to the ten thousands place, it would have
a value of _____.

If this digit were to move to the hundred thousands place, it would
have a value of _____.

If this digit were to move to the millions place, it would have
a value of _____.

The digit in each place has a value that is ten times as
great as it has in the place to its right.

Guided Practice

Circle the correct place of the highlighted digit and write
its value.

	Place		Value
1. 62,574	ones	tens	_____
2. 53,456	ten thousands	thousands	_____
3. 59,833	tens	hundreds	_____
4. 174,305	ten thousands	thousands	_____

Talk MATH

How does the value of a
digit in the thousands place
compare to its value if the
same digit was in the
hundreds place?

Independent Practice

Circle the place of the highlighted digit and write its value.

		Place		**Value**
5. 593,802	tens		hundreds	_____
6. 4,826,193	ten thousands		hundred thousands	_____
7. 7,830,259	hundred thousands		millions	

Use the place-value chart for Exercises 8–16.

Thousands Period			Ones Period		
hundreds	tens	ones	hundreds	tens	ones
4	6	2	3	7	1

8. The 6 is in the _____ place.

9. The _____ is in the thousands place.

10. The 7 has a value of 7 × _____.

11. The 6 has a value of 6 × _____.

12. The _____ has a value of _____ × 100,000.

13. The _____ is in the hundreds place.

14. The 1 is in the _____ place.

15. The digit in the hundred thousands place has 10 times the value

it would have if it was in the _____ place.

16. The digit in the thousands place has _____ times the
value it would have if it was in the hundreds place.

Problem Solving

17. An African elephant can weigh up to 14,432 pounds. What is the value of the highlighted digit?

18. Processes &Practices **2** **Stop and Reflect** Use the clues below to find the distance, in miles, from Earth to the Moon. Write the number in the place-value chart.

- The greatest place value is hundred thousands.
- The digit in the tens place is 5.
- The remaining digits are 2, 3, 8, and 7.
- One of the digits has a value of 30,000.
- One of the digits has a value of 800. The value of the digit in the thousands place is 10 times greater than this.
- There are 2 more ones than tens.

Thousands Period			Ones Period		
hundreds	tens	ones	hundreds	tens	ones

What is the distance from Earth to the Moon?

Brain Builders

19. Processes &Practices **4** **Model Math** What are the greatest and the least 6-digit numbers that have a 9 in the hundreds place and 3 fewer hundred thousands than hundreds?

20. Building on the Essential Question How does moving the place of a digit change its value? Give an example and explain.

Name ....

Homework Helper

Need help? connectED.mcgraw-hill.com

Write the place and value of the highlighted digit in 8,304,421.

Use a place-value chart.

Millions Period			Thousands Period			Ones Period		
hundreds	tens	ones	hundreds	tens	ones	hundreds	tens	ones
		8	3	0	4	4	2	1

The **3** is in the hundred thousands place.

The value of the **3** is 3 × 100,000, or 300,000.

Practice

Circle the place of the highlighted digit and write its value.

		Place		Value
1. 62,468	thousands		ten thousands	_____
2. 934,218	thousands		ten thousands	_____
3. 438,112	ten thousands		hundred thousands	_____
4. 285,012	tens		thousands	_____
5. 2,905,146	hundred thousands		millions	_____
6. 6,034,215	ten thousands		millions	_____

Problem Solving

Millions Period			Thousands Period			Ones Period		
hundreds	tens	ones	hundreds	tens	ones	hundreds	tens	ones

7. Write 1 in the hundreds place.

8. Write 8 in the tens place and 4 in the ones place.

9. Write 2 in the hundred thousands place.

10. Write 3 in the thousands place.

11. Write 7 in the millions place.

Brain Builders

12. Write 0 in the ten thousands place. What is its value? Explain how
    you know.

13. Write a number that is six hundred thousand more than the number.

Vocabulary Check

Match each definition to the correct vocabulary term.

14. The value given to a digit by its position in a number.   • digits

15. Symbols used to write whole numbers.   • place value

16. **Test Practice** A digit is in the hundreds place. The digit is moved so that its
    value is ten times greater. To which place did the digit move?

    Ⓐ hundred thousands   Ⓒ thousands

    Ⓑ ten thousands   Ⓓ tens

Name

....

ESSENTIAL QUESTION
How does place value help represent the value of numbers?

Place-value charts show the value of each digit. A group of three digits is called a **period**. Commas separate the periods. At each comma, say the name of the period.

Math in My World   Tools   Watch

Example 1

Scientists found that a gooney bird once traveled 24,983 miles in just 90 days. Use a place-value chart to read the number of miles that the gooney bird traveled.

The place-value chart shows 24,983.

Thousands Period			Ones Period		
hundreds	tens	ones	hundreds	tens	ones
	2	4 ,	9	8	3

The comma is after the thousands period. Say *thousand* when reading the number.

Say: *twenty-four* thousand, *nine hundred eighty-three*

So, the gooney bird traveled *twenty-four thousand, nine hundred eighty-three* miles.

There are different ways to write numbers. **Standard form**, or number names, only uses digits to show the number. **Expanded form** shows the number as a sum of the values of each digit. **Word form** shows the number only using words.

Example 2

The population of Botswana is about 1,882,000. Write this number in standard form, expanded form, and word form.

The place-value chart shows the place of each digit.

Millions Period			Thousands Period			Ones Period		
hundreds	tens	ones	hundreds	tens	ones	hundreds	tens	ones
		1	8	8	2	0	0	0

Write the number in standard form and expanded form.

☐ , 8 ☐ ☐ , 0 ☐ ☐ ← (standard form)

1,000,000 + _____ + 80,000 + _____ ← (expanded form)

Write the number in word form.

one million, _____ hundred eighty-_____ thousand

Guided Practice

Write each number in standard form.

1. *three hundred forty-nine thousand, twenty-five* _____

2. 400,000 + 90,000 + 2,000 + 800 + 10 + 4 _____

Write each number in expanded form and word form.

3. 492,032

_____

_____

4. 3,028,002

_____

_____

What is the value of the 6 in 345,629?

Name ....

Write each number in standard form.

5. *twenty-five thousand, four hundred eight* _____

6. *forty thousand, eight hundred eleven* _____

7. *seven hundred sixty-one thousand, three hundred fifty-six* _____

8. *five million, seven hundred sixty-two thousand, one hundred eleven*

   _____

9. 600,000 + 80,000 + 4 _____

10. 20,000 + 900 + 70 + 6 _____

11. 9,000,000 + 200,000 + 1,000 + 500 + 2 _____

Write each number in expanded form and word form.

12. 485,830

    Expanded Form:

    _____

    Word Form:

    _____

    _____

13. 3,029,251

    Expanded Form:

    _____

    Word Form:

    _____

Problem Solving

14. **Processes &Practices** 1 **Plan Your Solution** A zoo's newborn African elephant weighed 232 pounds. After one year, the elephant gained 1,000 pounds. Write the elephant's new weight in expanded form and word form.

Expanded form: _____

Word Form: _____

_____

15. The population of Norway is about *four million, seven hundred two thousand*. Write this number in standard form.

_____

Brain Builders

16. The population of the Dominican Republic is about 8,664,000 more than that of Norway. Write this number in word form.

_____

_____

17. **Processes &Practices** 3 **Find the Error** Sonia wrote the expanded form of 2,408,615 below.

$$2{,}000{,}000 + 400{,}000 + 80{,}000 + 600 + 10 + 5$$

Find and correct her mistake.

_____

_____

18. **Building on the Essential Question** Why is expanded form important? Explain.

_____

Name _____

Homework Helper

Need help? connectED.mcgraw-hill.com

Write 1,000,000 + 300,000 + 60,000 + 300 + 10 + 5
in standard form. Then read the number aloud.

standard form: 1,360,315

Remember: Commas separate the periods. Say the name
of the period at each comma.

Millions Period			Thousands Period			Ones Period		
hundreds	tens	ones	hundreds	tens	ones	hundreds	tens	ones
		1,	3	6	0,	3	1	5

Say: *one million, three hundred sixty thousand, three hundred fifteen*

Write 756,491 in expanded form and word form.

expanded form: 700,000 + 50,000 + 6,000 + 400 + 90 + 1

word form: *seven hundred fifty-six thousand, four hundred ninety-one*

Practice

1. Write *one million, one hundred forty-five thousand,*
   two hundred thirty-seven in standard form.

   _____

2. Write 87,192 in word form and expanded form.

   _____

   _____

Problem Solving

Complete the expanded form.

3. **Processes &Practices** **1** **Check for Reasonableness**

91,765 = 90,000 + _____ + 700 + _____ + 5

4. 798,054 = 700,000 + _____ + _____ + 50 + 4

5. 5,925,020 = 5,000,000 + _____ + 20,000 + _____ + 20

Brain Builders

6. Explain how to write the expanded form when a number includes a zero.

_____

_____

Vocabulary Check

Read each definition. Choose the correct word(s) to fill in the spaces.

expanded form   period

standard form   word form

7. the way of writing a number using words

_____ _____ _____ _____ _____

8. the usual way of writing a number, using digits

_____ _____ _____ _____ _____ _____ _____

9. the way of writing a number as the sum of the value of each digit

_____ _____ _____ _____ _____ _____ _____

10. each group of three digits on a place-value chart

_____ _____ _____ _____ _____

11. **Test Practice** Which is the correct expanded form for 45,098?

  Ⓐ 45,000 + 98   Ⓒ 40,000 + 500 + 90 + 8

  Ⓑ 4,000 + 5,000 + 9 + 8   Ⓓ 40,000 + 5,000 + 90 + 8

Name _____

Lesson 3
Compare Numbers

ESSENTIAL QUESTION
How does place value help represent the value of numbers?

A **number line** is a line with numbers on it in order at regular intervals. You can use a number line to compare numbers. Use these symbols to show how numbers relate to each other.

| is greater than (>) | is less than (<) | is equal to (=) |

 Math in My World Tools Watch Tutor

Example 1

On average, a first-year police officer earns $41,793. A first-year firefighter earns $41,294. Which occupation pays more the first year?

Label each dot with the correct salary.

41,000   41,200   41,400   41,600   41,800   42,000

← Numbers get smaller.   Numbers get larger. →

41,793 is to the

_____ of 41,294.

41,793 is _____ than 41,294.

41,793 ◯ 41,294   ← Write the > symbol in the circle.

So, _____ earn more the first year than _____ .

Copyright © McGraw-Hill Education ©FangXiaNuo/iStock/Getty Images

Online Content at connectED.mcgraw-hill.com

Lesson 3 Compare Numbers **23**

Example 2

In a recent year, the population of Vermont was **621,760**. The population of North Dakota was **646,844**. Compare these two populations. Use <, >, or =.

Use a place-value chart.

Thousands Period			Ones Period		
hundreds	tens	ones	hundreds	tens	ones
			,		
			,		

same   different

2 < 4

1 Write the numbers on the place-value chart.

2 Compare the digits of the greatest place value. If they are the same, move to the next digit to the right until you find digits that are different.

So, 621,760 ◯ 646,844.

Guided Practice

1. Use the number line to compare. Use <, >, or =.

32,053 ◯ 35,251

30,000 31,000 32,000 33,000 34,000 35,000 36,000

Compare. Use <, >, or =.

2. 25,409 ◯ 26,409

3. 13,636 ◯ 13,636

4. 72,451 ◯ 76,321

5. 201,953 ◯ 201,953

6. 442,089 ◯ 442,078

7. 224,747 ◯ 224,774

Talk MATH

If two numbers have all of the same digits in the same places, can one of them be greater than the other? Explain.

Name ....

**For Exercises 8–10, use the number lines to compare.
Use <, >, or =.**

8. 45,526 ◯ 48,873

43,000 44,000 45,000 46,000 47,000 48,000 49,000

9. 31,748 ◯ 31,521

31,300 31,400 31,500 31,600 31,700 31,800 31,900

10. 126,532 ◯ 129,321

125,000 126,000 127,000 128,000 129,000 130,000 131,000

Compare. Use <, >, or =.

11. 3,030 ◯ 3,030

12. 76,101 ◯ 77,000

13. 12,683 ◯ 12,638

14. 229,214 ◯ 300,142

15. 701,000 ◯ 701,000

16. 342,646 ◯ 34,646

17. 398,421 ◯ 389,421

18. 605,310 ◯ 605,310

19. 840,515 ◯ 845,015

20. 655,543 ◯ 556,543

21. 720,301 ◯ 720,031

22. 333,452 ◯ 333,452

Problem Solving

23. Jun collects stamps and baseball cards. He has 1,834 stamps and 1,286 baseball cards. Does he have more stamps or more cards? Explain.

_____

24. The population of Cindy's city is 242,506. The population of Mark's city is 242,605. Who lives in the city with the greater population? Explain.

_____

Brain Builders

25. **Processes &Practices** 6 **Explain to a Friend** Explain how to compare numbers using place value. Provide an example.

_____

_____

Processes &Practices 2 **Understand Symbols** For Exercises 26–28, fill in each blank to make the number sentence true. How did you chose each number?

26. 253,052 < _____

_____

27. 95,925 > _____

_____

28. 205,053 < _____

_____

29. **Building on the Essential Question** How can I show how numbers are related to each other?

_____

_____

Name ....

MY Homework

Homework Helper Need help? connectED.mcgraw-hill.com

Compare 54,515 and 54,233. Use >, <, or =.

Use a number line.

54,515 is to the right of 54,233 on the number line.

So, 54,515 > 54,233.

Practice

For Exercises 1–2, use the number lines to compare. Use <, >, or =.

1. 67,113 ◯ 62,523

2. 42,254 ◯ 42,533

Compare. Use <, >, or =.

3. $751,012 ◯ $715,012   4. 4,350 ◯ 5,430   5. 8,080 ◯ 8,880

6. 322,650 ◯ 332,650   7. 673 ◯ 376   8. $918,050 ◯ $819,050

9. 121,571 ◯ 211,571   10. 17,888 ◯ 17,780   11. 72,770 ◯ 72,770

Problem Solving

Processes
12. &Practices **3** **Draw a Conclusion** Myron has $1,698 in his savings account. Robert has $1,898 in his savings account. Toby has $100 less than Robert in his savings account. Who has the least amount of money?

_____

Brain Builders

13. There were 544,692 tickets sold for the rock concert. There were 455,692 tickets sold for the country music concert. Which concert sold a greater number of tickets? Explain how you know.

_____

_____

Vocabulary Check

14. Choose the correct word(s) to complete each sentence.

is equal to (=)   is greater than (>)
is less than (<)   number line

To compare numbers, you can use a _____. A number that is to the right on a number line _____ a number to its left. A number on the left _____ a number to its right.

You can look at place values to compare numbers. If a number has a digit in the thousands place that _____ the thousands digit in another number, then look to the hundreds place.

15. Test Practice Which number sentence is *not* true?

Ⓐ 243,053 < 242,553   Ⓒ 553,025 = 553,025

Ⓑ 194,832 > 193,832   Ⓓ 295,925 < 295,952

Name ....

Lesson 4
Order Numbers

ESSENTIAL QUESTION
How does place value help represent the value of numbers?

You can use a place-value chart to put numbers in order.

 Math in My World Tutor

Lowell pop. 103,299

Cambridge pop. 101,365

Boston pop. 590,763

Example 1

Compare the populations of the three cities and order them from *greatest* to *least*.

1 Write the populations on the place-value chart.

2 Start with the greatest place-value position. Compare.

_____ > _____

3 Compare the digits in the next place.

_____ = _____

4 Continue to compare until the digits are different.

_____ > _____

_____ > _____ > _____

So, from *greatest* population to *least*, the order of the cities is

_____, _____, and _____.

Thousands			Ones		
hundreds	tens	ones	hundreds	tens	ones

Copyright © McGraw-Hill Education

Online Content at connectED.mcgraw-hill.com

Lesson 4 Order Numbers 29

Example 2 Tutor

Order the numbers on the cards from _least_ to _greatest_.

1 Line up the numbers by the ones place.

☐ ☐ ☐ , ☐ ☐ ☐

☐ ☐ ☐ , ☐ ☐ ☐

☐ ☐ ☐ , ☐ ☐ ☐

245,032   254,002

245,023

2 Start with the greatest place-value position. Compare.

Each number has a _____ in the hundred

thousands place. So, compare the digits in the

ten thousands place. The greatest number is _____ .

The two numbers that are left both have a _____ in

the thousands place and a _____ in the hundreds place.

The second greatest number is _____ .

So, in order from _least_ to _greatest_, the numbers are

_____ .

Guided Practice

1. Place the numbers in the place-value chart in order from _greatest_ to _least_.

   52,482
   50,023
   56,028
   63,340

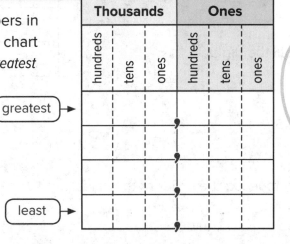

Thousands			Ones		
hundreds	tens	ones	hundreds	tens	ones
greatest →		,			
		,			
		,			
least →		,			

Talk MATH

When ordering numbers, what do you do when the digits in the same place have the same value?

30 Chapter 1 Place Value

Copyright © McGraw-Hill Education

Name

Place the numbers in the place-value chart in order from *greatest* to *least*.

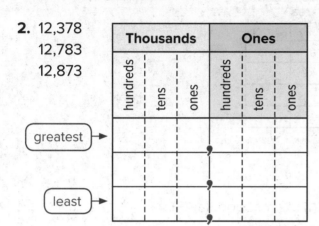

2. 12,378
   12,783
   12,873

3. 258,103
   248,034
   285,091
   248,934

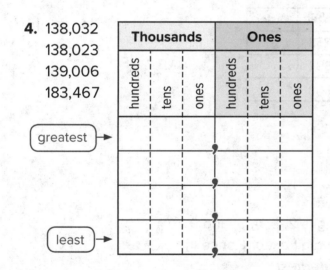

4. 138,032
   138,023
   139,006
   183,467

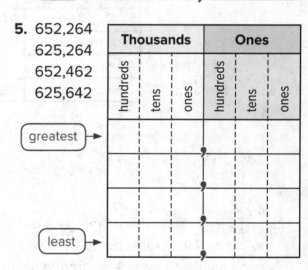

5. 652,264
   625,264
   652,462
   625,642

Order the numbers from *least* to *greatest*.

6. 402,052; 425,674; 414,035

7. 643,947; 643,537; 642,066

8. 113,636; 372,257; 337,633

9. 563,426; 564,376; 653,363

Problem Solving

10. **Processes &Practices** 2 **Stop and Reflect** Order the states from *least* (1) to *greatest* (4) total area.

Land and Water Area

State	Total Area (sq. mi)	Order
Wyoming	97,814	
Alaska	663,267	
Texas	268,581	
California	163,696	

11. Order the dog breeds from *least* popular (1) to *most* popular (3).

Dog Breeds

Dog Breed	Number	Order
Yorkshire Terrier	47,238	
Beagle	42,592	
German Shepherd	45,868	

Brain Builders

12. **Processes &Practices** 1 **Keep Trying** Use the digits 2, 3, 4, 5, and 9 to create four 5-digit numbers. Use each digit exactly once in each number. Order them from *least* to *greatest*.

_____

_____

13. **Building on the Essential Question** When do I compare real-world numbers?

_____

_____

Name ....

Homework Helper

Need help? connectED.mcgraw-hill.com

Order the numbers from *greatest* to *least*:
17,601; 20,007; 17,610

Compare the ten thousands.

17,601
20,007 ◄——— most ten thousands
17,610

Both thousands and hundreds are the same,
so compare the tens.

17,601
17,610 ◄——— more tens

From *greatest* to *least*, the numbers are 20,007; 17,610; and 17,601.

Practice

Order the numbers from *greatest* to *least*.

1. 59,909; 95,509; 59,919   **2.** 2,993; 9,239; 2,393

3. 112,443; 114,324; 112,344   **4.** 642,063; 642,036; 642,306

Order the numbers from *least* to *greatest*.

5. 225,625; 335,432; 325,745   **6.** 357,925; 329,053; 356,035

Problem Solving

7. The United States' soccer team has 572,112 fans. Great Britain's team has 612,006 fans. Brazil's team has 901,808 fans. Write the countries in order from the *greatest* to *least* number of soccer fans.

_____

_____

8. There are 943,025 sports tickets available. There are 832,502 movie tickets available. There are 415,935 theater tickets available. List the number of tickets in order from *least* to *greatest*.

_____

_____

Brain Builders

9. **Processes &Practices** **7** **Identify Structure** Write four numbers that each have six digits. Order the numbers from *least* to *greatest*.

_____

_____

10. **Test Practice** The table shows the populations of the cities where Alex and Brent live. Marcia lives in a city that has more people than Alex's city and fewer people than Brent's. Which could be the number of people who live in Marcia's city?

Name	Population of their city
Alex	404,048
Brent	412,888

Ⓐ 413,066 people   Ⓒ 404,132 people

Ⓑ 412,901 people   Ⓓ 403,997 people

Check My Progress

Vocabulary Check

1. Each card shows the definition or an example of a vocabulary word. Write each word from the word bank on the card with the matching definition or example.

digit	expanded form	is equal to (=)	is greater than (>)
is less than (<)	number line	period	place value
standard form	word form		

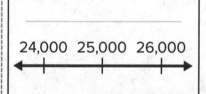

24,000 25,000 26,000

_____

23,042 ◯ 23,000

_____

Example: 83,104

_____

the value given to a digit by its position in a number

_____

any symbol used to write a whole number

_____

the name given to each group of three digits on a place-value chart

_____

Example:
80,000 + 3,000 + 100 + 4

_____

34,842 ◯ 43,842

_____

Example:
eighty-three thousand, one hundred four

_____

44,204 ◯ 44,204

Concept Check

Write the place of the highlighted digit. Then write its value.

2. 34,025

3. 52,276

 Problem Solving

4. A car costs *thirty-six thousand, five hundred forty-seven* dollars. Write the cost in standard form.

 Brain Builders

5. A farm sold 429,842 apples and 53,744 pears. Did the farm sell more apples or pears? Explain.

6. The table shows how much money was made at Laser Flip and Grind's Skateboard Store.

Laser Flip and Grind's Skateboard Store Sales	
Item	**Sales**
Skateboards	$130,223
Helmets	$103,322
Ramps	$133,032

Explain and show how to write the sales in order from *greatest* to *least*.

7. Test Practice Which set of numbers is written from *least* to *greatest*?

Ⓐ 351,935; 351,914; 215,634

Ⓑ 351,914; 215,634; 351,935

Ⓒ 215,634; 351,935; 351,914

Ⓓ 215,634; 351,914; 351,935

Name

Lesson 5
Use Place Value to Round

ESSENTIAL QUESTION
How does place value help represent the value of numbers?

When you estimate, you find an answer that is close to the exact answer. One way to estimate is to round by changing the value of a number so it is easier to work with.

 Math in My World   Tools   Watch   Tutor

Example 1

The largest extreme sports competition, called X-Games, is so popular that one year 268,390 people attended. What is 268,390 rounded to the nearest ten thousand?

Look at the number line. 268,390 is between 260,000 and 270,000.

268,390

260,000   262,000   264,000   266,000   268,000   270,000

Since 268,390 is closer to _____ than

_____ , 268,390 rounds to _____ .

Copyright © McGraw-Hill Education   ©John Kirkpatrick/Alamy

Example 2

A record was set when 569,069 people jumped up and down for one minute. About how many people set this record? Round 569,069 to the nearest hundred thousand.

 Circle the digit in the place to be rounded.

569,069

2 Underline the digit to the right of the place being rounded.

3 If the underlined digit is 4 or less, do not change the circled digit. If the underlined digit is 5 or greater, add 1 to the circled digit.

4 Replace all digits after the circled digit with zeros.

So, about _____ people set this record.

Check

Use a number line. 569,069 is closer to _____ than 500,000.

569,069

500,000   550,000   600,000

Guided Practice

Round to the given place-value position.

1. 2,221; thousands _____

2. 78,214; ten thousands _____

3. 581,203; hundred thousands _____

What is the least number that you can round to the thousands place to get 8,000? Explain.

Independent Practice

Round to the given place-value position.

4. 500,580; thousands

_____

5. 290,152; hundred thousands

_____

6. 218,457; hundred thousands

_____

7. 37,890; hundreds

_____

8. 95,010; thousands

_____

9. 845,636; ten thousands

_____

10. 336,001; hundred thousands

_____

11. 709,385; hundred thousands

_____

Tell the place-value position to which each number was rounded.

12. 456,750 ⟶ 460,000

_____

13. 38,124 ⟶ 38,120

_____

14. 18,334 ⟶ 18,000

_____

15. 455,670 ⟶ 455,700

_____

16. 980,065 ⟶ 980,070

_____

17. 162,245 ⟶ 200,000

_____

Problem Solving

A natural gas and fuel-friendly car was used to set a world record.

World Record
23,697 miles or
38,137 kilometers

18. What is the distance in miles traveled to the nearest ten thousand?

_____

19. Round the distance traveled in kilometers to the nearest thousand.

_____

20. The largest house made out of playing cards used 91,800 cards. To the nearest thousand, how many cards were used?

_____

Brain Builders

21. **Processes &Practices** ⑥ **Explain to a Friend** Earth's deepest point is the Mariana Trench in the Pacific Ocean. It is 35,840 feet below sea level. Would it be more precise to round the depth to the nearest ten thousand or the nearest hundred? Explain.

_____

_____

22. **Processes &Practices** ③ **Find the Error** Andrew rounded the number 672,726 to the nearest hundred thousand. He wrote 672,000. Find and correct his mistake.

_____

_____

23. ❓ **Building on the Essential Question** When is estimation an effective way to determine an answer?

_____

Name ....

 MY Homework

Homework Helper

Need help? ⇗ connectED.mcgraw-hill.com

Round 65,839 to the nearest hundred.

Circle the digit to be rounded.   65,⑧39

The digit to the right is 4 or less, so the 8 does not change. All digits after the 8 are replaced with zeros.

65,839 rounded to the nearest hundred is 65,800.

Round 65,839 to the nearest ten thousand.

Circle the digit to be rounded.   ⑥5,839

The digit to the right is 5 or more, so 1 is added to the circled digit. The digits after the circled digit are replaced with zeros.

65,839 rounded to the nearest ten thousand is 70,000.

Practice

Round each number to the given place-value position.

1. 64,569; thousands

2. 155,016; thousands

3. 73,569; ten thousands

4. 708,569; ten thousands

5. 91,284; hundred thousands

6. 265,409; hundred thousands

Copyright © McGraw-Hill Education

Problem Solving

7. Luis and his family flew 51,487 miles last summer while on vacation. Rounded to the nearest thousand, how many miles did they fly?

8. Miles bought a car that cost $23,556. To the nearest ten thousand, how much did the car cost?

Brain Builders

9. Explain how you would round the numbers 33 and 89 to estimate their sum.

Use the data from the table for Exercises 10–12.

Depths of Oceans	
Ocean	Average Depth (ft)
Pacific	12,925
Atlantic	11,730
Indian	12,598

10. **Processes &Practices** **5** **Use Math Tools** Which ocean has an average depth of about 12,000 feet, to the nearest thousand? How do you know?

11. The depth of which two oceans round to the same thousand?

12. What is the depth of the Indian Ocean rounded to the nearest thousand? Explain.

13. **Test Practice** What is 104,229 rounded to the nearest ten thousand?

Ⓐ 90,000 Ⓑ 100,000 Ⓒ 104,000 Ⓓ 110,000

Name _____

Lesson 6
Problem-Solving Investigation
STRATEGY: Use the Four-Step Plan

ESSENTIAL QUESTION
How does place value help represent the value of numbers?

Learn the Strategy

Ben, Andy, and Kelly each live in a different city. The populations of the cities are 372,952; 225,395; and 373,926. Use the clues to find the population of the city where Ben lives.

Clues

- Andy's city has the least population.
- When rounded to the nearest thousand, the population of Kelly's city is 374,000.

1 Understand

What facts do you know?

Ben, Andy, and Kelly each live in a different city.

The populations for each city are: _____; _____; and _____.

What do you need to find?

the population of _____ city

2 Plan

I can order and round the populations.

3 Solve

Order the populations from *least* to *greatest*. 225,395; 372,952; 373,926

_____ lives in the city with the least population.

Round the remaining populations to the nearest thousand.

372,952 rounds to _____. 373,926 rounds to _____.

Kelly lives in the city with the population that rounds to 374,000.

So, Ben must live in the city that has a population of _____.

4 Check

Does your answer make sense? Explain.

_____

It is estimated that a movie made more than $7,000,000 but less than $8,000,000. There is a 5 in the thousands place, a 7 in the ten thousands place, and a 6 in the hundred thousands place. The ones, tens, and hundreds places have zeros in them because the total is an estimate. What is the estimated amount of money that the movie made?

 Understand

What facts do you know?

_____

_____

What do you need to find?

_____

_____

 Plan

 Solve

Check

Does your answer make sense? Explain.

_____

Apply the Strategy

Solve each problem by using the four-step plan.

1. **Processes &Practices** 5 **Use Math Tools** Mr. Kramer is buying a car. The list of prices is shown in the table.

   Mr. Kramer wants to buy the least expensive car.

   Which car should he buy?

Prices of Cars	
Cars	Price
Car A	$83,532
Car B	$24,375
Car C	$24,053
Car D	$73,295

2. A restaurant made more than $80,000 but less than $90,000 last month. There is a 6 in the ones place, a 3 in the thousands place, a 7 in the hundreds place, and a 1 in the tens place. How much money did the restaurant make last month?

Brain Builders

3. Amy, Lisa, Angie, and Doug live in different states. The populations of those states are 885,122; 5,024,748; 4,492,076; and 2,951,996. Lisa lives in the state with the greatest population. Doug lives in the state with a 2 in the thousands place of the population. Amy lives in the state with the least population. What is the population of Angie's state?

4. The New Meadowlands Stadium in New Jersey seats a large number of fans. There are zeros in the tens and ones places, and a 5 in the hundreds place. The number in the thousands place is 2 less than the hundreds place. The number in the ten thousands place is 3 more than the number in the hundreds place. How many people does the stadium seat?

Review the Strategies

Use any strategy to solve each problem.

- Make a table.
- Choose an operation.
- Act it out.
- Draw a picture.

5. The table at the right shows which types of canned foods were collected during a food drive.

What was the most popular type of canned food collected?

Canned Foods	Number Collected
tomatoes	59,294
beans	159,002
corn	45,925
soup	903,690

6. The population of a city has 6 digits. There is a 3 in the tens place, a 5 in the hundred thousands place, a 6 in the ones place, and a 9 in the rest of the places. What is the population of the city?

7. A warehouse stores cans of paint. There is a 3 in the hundreds place, a 7 in the thousands place, a 5 in the ten thousands place, and an 8 in the rest of the places. This number has 5 digits. How many cans of paint are in the warehouse?

Processes
8. **&Practices** **7** **Identify Structure** A car's mileage is a five-digit number. There is a 3 in the ten thousands place, the ones place, and the tens place. There is a 9 in the hundreds place and the thousands place. What is the car's mileage?

Name ....

MY Homework

Homework Helper eHelp

Need help? ⟋ connectED.mcgraw-hill.com

A six-digit number has a 2 in the thousands place, a 5 in the tens place, a 3 in the hundred thousands place, and zeros in each of the remaining places. What is the number?

Use the four-step plan to solve this problem.

 Understand

I know that there is a number with six digits. It has a 2 in the thousands place, a 5 in the tens place, a 3 in the hundred thousands place, and zeros in each of the remaining places.

I need to find the number.

 Plan

I will use a place-value chart to help me organize the digits.

 Solve

Thousands Period			Ones Period		
hundreds	tens	ones	hundreds	tens	ones
3	0	2	0	5	0

So, the number is 302,050.

 Check

I can check my work by reading the clues again to make sure that the digits are all in the correct places.

Problem Solving

1. A five-digit number has a 3 in the hundreds place, a 7 in the greatest place-value position, a 9 in the ones place, an 8 in the thousands place, and a 6 in the tens place. What is the number? Use the four-step plan.

Solve each problem by using the four-step plan.

2. Use the digits 1–7 to create a seven-digit number that can be rounded to 6,300,000.

_____

3. A seven-digit number has a 0 in the ones place, a 6 in the ten thousands place, an 8 in the millions place, and fives in each of the remaining places. What is the number?

_____

Brain Builders

4. Tara rolled the numbers shown. What is the greatest number she can make using every digit once? What is the second greatest number?

_____

5. Betsy, Carl, and Dave each live in different cities. The populations of the cities are 194,032; 23,853; and 192,034. Betsy lives in the city with the least population. Carl does not live in the city with the greatest population. What is the population of Dave's city?

_____

6. **Processes &Practices** **6** **Explain to a Friend** Explain how the value of the 7 in 327,902 will change if you move it to the tens place.

_____

_____

Review

Vocabulary Check

Use the words in the word bank to complete each sentence.

digits   expanded form   is equal to (=)

is greater than (>)   is less than (<)   number line

period   place value   standard form

word form

1. 83,502 _____ 82,502.

2. You can use a _____ to compare numbers.

3. There are five _____ in the number 35,024.

4. 392,903 _____ 392,903.

5. The _____ of 32,052 is *thirty-two thousand, fifty-two.*

6. The _____ of 853,025 is 800,000 + 50,000 + 3,000 + 20 + 5.

7. _____ is the value given to a digit by its position in a number.

8. The name given to each group of three digits on a place-value chart is called a _____ .

9. The _____ of *fifteen thousand, sixty-two* is 15,062.

10. 473,503 _____ 474,503.

Concept Check

11. Write *two hundred thirty-nine thousand, eight hundred four* in standard form and expanded form.

_____

_____

Compare. Use <, >, or =.

12. 689,000 ◯ 679,000

13. 515,063 ◯ 515,603

14. 739,023 ◯ 739,023

15. 405,032 ◯ 450,002

16. Round 415,203 to the thousands place. _____

Use the place-value chart for Exercises 17–23.

Thousands Period			Ones Period		
hundreds	tens	ones	hundreds	tens	ones
5	3	7	2	8	0

17. The 3 is in the _____ place.

18. The _____ is in the thousands place.

19. The 8 has a value of 8 × _____ .

20. The 3 has a value of 3 × _____ .

21. The _____ has a value of _____ × 100,000.

22. The _____ is in the hundreds place.

23. The digit in each place has a value that is _____ times as great as it has in the place to its _____ .

Order the numbers from *greatest* to *least*.

24. 374,273; 374,372; 347,732

_____

25. 263,224; 623,224; 633,222

_____

Problem Solving

26. There were 48,566 people at a Sunday football game. To the nearest thousand, how many people were at the game?

27. The table shows the cost of three houses. Order these prices from least to greatest.

House	Price
House A	$175,359
House B	$169,499
House C	$179,450

Brain Builders

28. In a recent year, the population of Hong Kong was about 6,924,000. What is the value of the 9 in this number? Write another 7-digit number with 9 having the same value.

29. A baseball stadium has 24,053 seats. A soccer stadium has 20,000 more seats than the baseball stadium. A football stadium has 53,025 seats. Order the number of seats from *greatest* to *least*.

30. **Test Practice** The population of New Zealand is about 4,184,000. What is the expanded form of this number rounded to the nearest ten thousand?

Ⓐ 4,000,000 + 100,000 + 80,000

Ⓑ 4,000,000 + 100,000 + 8,000

Ⓒ 400,000 + 10,000

Ⓓ 4 + 1 + 8 + 4

Reflect

Use what you learned about place value to
complete the graphic organizer.

Write the Example	Real-World Example
Vocabulary	Estimate

ESSENTIAL QUESTION

How does place value
help represent the
value of numbers?

Reflect on the ESSENTIAL QUESTION Write your answer below.

Performance Task

Brain Builders

Visiting Grand Teton National Park

Liam is doing a school project on the Grand Teton National Park. He also hopes to someday visit the park with his family.

Show all your work to receive full credit.

Part A

The table shows the number of visitors that the Grand Teton National Park has had during four months of the year. Complete the bar graph below the table to show the number of visitors for each month rounded to the nearest thousand.

Month	Number of Visitors to Grand Teton Park
April	21,847
May	25,934
June	37,016
July	35,335

Part B

For the rest of the year, between 400,000 and 500,000 people visited Grand Teton National Park. In the total number of visitors for the rest of the year, the digit 7 has 10 times the value that it has in the total visitors for June. In the total visitors in the rest of the year, the tens digit and the thousands digit are both the same even number, and the ones digit is three times the tens digit. The only other digit is a 5, and it is the only 5 in the number. How many people visited Grand Teton National Park for the rest of the year? Explain your reasoning.

Part C

Grand Teton National Park covers 1,254 square kilometers of land. Write the expanded form and the word form for Grand Teton's land area.

Part D

The table shows the heights of several mountains in the park.

Mountain	Height (feet)
Grand Teton	13,776
Teewinot Mt.	12,326
Mount Moran	12,605
South Teton	12,513
Mount Owen	12,927
Middle Teton	12,805

Use inequality symbols to order the height of the mountains from **least** to **greatest**. Explain how you used place value to order the heights.

Chapter

2 Add and Subtract Whole Numbers

ESSENTIAL QUESTION

What strategies can I use to add or subtract?

Let's Watch the Show!

Watch

Watch a video!

Brain Builders

MY Chapter Project

Recycle It!

1. Your class will have a drive to collect cans and bottles. Work with your group to make a list of the steps to organize the drive. Include any questions your group may have. Example: *Who will take the cans and bottles to the recycle center?* Discuss your steps and questions with the class.

2. Your group will make a poster to promote your drive. As a class, create a list of things to include on the posters. Examples: *When will the drive start? When will it end?* Post a list in the classroom. As a group, create your poster.

3. Estimate how many cans and bottles your class will collect each week. Write your estimates in the table below. At the end of each week, write the actual number of cans and bottles your class collected.

	Estimate		Actual	
	Cans	Bottles	Cans	Bottles
Week 1				
Week 2				
Week 3				
Week 4				

Name ....

Am I Ready?

Add.

1. 35
    + 56

2. $58
    + $25

3. 94
    + 78

4. $87 + $35 = _____

5. 103 + 57 = _____

6. 233 + 158 = _____

7. Felicia has a collection of 117 marbles. Her
   sister gives her 25 marbles. How many marbles
   does Felicia have now?

   _____

Subtract.

8. $57
    − $8

9. 71
    − 23

10. 132
    − 74

11. 93 − 15 = _____

12. $62 − $49 = _____

13. 415 − 107 = _____

14. Jasper is reading a 98-page book. He has read 29 pages.
    How many pages does Jasper have left to read?

    _____

Shade the boxes to show the problems you answered correctly.

How Did I Do? | 1 | 2 | 3 | 4 | 5 | 6 | 7 | 8 | 9 | 10 | 11 | 12 | 13 | 14 |

Online Content at connectED.mcgraw-hill.com

Review Vocabulary

| difference | estimate | round | sum | word form |

Making Connections

Use the review vocabulary to describe the examples based on the problems in each chart.

Subtraction Problem	Vocabulary Word	Example
		$2,200 - 600 = 1,600$
2,238 − 599		*two thousand, two hundred thirty-eight* minus *five hundred ninety-nine*
		1,639
		$2,200 - 600 = 1,600$

Addition Problem	Vocabulary Word	Example
		$5,900 + 700 = 6,600$
5,877 + 673		*five thousand, eight hundred seventy-seven* plus *six hundred seventy-three*
		6,550
		$5,900 + 700 = 6,600$

MY Vocabulary Cards

Processes
&Practices

Copyright © McGraw-Hill Education

Lesson 2–1

Associative Property of Addition

$(13 + 10) + 4 = 13 + (10 + 4)$

Lesson 2–1

Commutative Property of Addition

$12 + 15 = 15 + 12$

Lesson 2–9

equation

$a + 2 = 5; 7 - b = 4$

Lesson 2–1

Identity Property of Addition

$0 + 18 = 18 \quad\mid\quad 18 + 0 = 18$

Lesson 2–6

minuend

$1,000 - 17 = 983$

Lesson 2–6

subtrahend

$1,000 - 17 = 983$

Lesson 2–1

unknown

$150 + 300 - 200 = ?$

Lesson 2–9

variable

$150 + 300 - k = 250$

Ideas for Use

- Draw examples for each card. Make drawings that are different from what is shown on each card.

- Practice your penmanship! Write each word in cursive.

The property which states that the order in which two numbers are *added* does not change the *sum*.

Imagine you need to shop for many items. Describe how this property could be used when estimating the sum.

_____

_____

The property which states that the grouping of the *addends* does not change the *sum*.

Associate can mean "to combine or come together." Explain how this helps you understand this property.

_____

_____

For any number, zero plus that number is the number.

Look in the dictionary for a meaning of *identity*. Write a sentence using that meaning.

_____

_____

_____

A sentence that contains an equals sign (=), showing that two expressions are equal.

Explain how an equation is different than an expression.

_____

_____

A number subtracted from another number.

Write a tip to help you remember which number is the minuend and which is the subtrahend.

_____

_____

The number from which another number is subtracted.

What word from this chapter is also a part of a subtraction equation, other than minuend?

_____

A letter or symbol used to represent an unknown quantity.

Variable can mean "changeable." Explain how this relates to the math meaning of *variable*.

_____

_____

An amount that has not been identified.

The prefix *un-* means "not." Name two other words with this prefix.

_____

MY Foldable

FOLDABLES® Follow the steps on the back to make your Foldable.

✂ -

Subtraction Equation	Check with Addition
$2{,}161 - 125 = 2{,}036$	$2{,}036 + 125 = 2{,}161$
$7{,}013 - 1{,}692 = 5{,}321$	$5{,}321 + 1{,}692 = \underline{\quad}$
$\underline{\quad} - \underline{\quad} = \underline{\quad}$	$\underline{\quad} + \underline{\quad} = \underline{\quad}$

Study Organizer

Addition Equation	Check with Subtraction
$2,036 + 125 = 2,161$	$2,161 - 125 = 2,036$
$5,321 + 1,692 = 7,013$	$7,013 - 1,692 = $ ____
___ $+$ ___ $=$ ___	___ $-$ ___ $=$ ___

Name ....

Lesson 1
Addition Properties and Subtraction Rules

ESSENTIAL QUESTION
What strategies can I use to add or subtract?

Addition properties can be used to help solve addition problems.

 Math in My World Watch ▶ Tutor 💬

Example 1

Carlos is buying the items shown. Does the order in which the musical instruments are scanned change the total cost?

$10 + $20 = $20 + $10

$ [   ] = $ [   ]

The order in which the instruments are scanned does not change the total cost. This is the Commutative Property of Addition.

Key Concept Addition Properties

Words	**Commutative Property of Addition** The order in which numbers are added does not change the sum.
Examples	4 + 1 = 5 1 + 4 = 5
Words	**Associative Property of Addition** The way in which numbers are grouped when added does not change the sum.
Examples	(5 + 2) + 3 5 + (2 + 3)
Words	**Identity Property of Addition** The sum of any number and 0 is the number.
Examples	8 + 0 = 8 0 + 8 = 8

(5 + 2) + 3 → 7 + 3 → 10

5 + (2 + 3) → 5 + 5 → 10

Parentheses ( ) show which numbers are added first.

Example 2

There were 16 people at the pool on Saturday. There were no people at the pool on Sunday. How many people were there on Saturday and Sunday?

_____ + _____ = _____ This is the _____
Property of Addition.

So, there were _____ people at the pool on Saturday and Sunday.

You can use properties and rules to find the **unknown**, or missing number, in a number sentence.

Example 3

Find the unknown in $10 - \blacksquare = 10$.

When you subtract 0 from any number, the result is the number.

So, the unknown is _____ .

Key Concept Subtraction Rules

Words	When you subtract 0 from any number, the result is the number.
Examples	$22 - 0 = 22$ $14 - 0 = 14$
Words	When you subtract any number from itself, the result is 0.
Examples	$16 - 16 = 0$ $20 - 20 = 0$

Guided Practice

Find each unknown. Draw a line to identify the property or rule used.

1. $19 - \blacksquare = 19$

   $\blacksquare =$ _____

2. $(5 + \blacksquare) + 2 = 5 + (9 + 2)$

   $\blacksquare =$ _____

3. $74 + 68 = \blacksquare + 74$

   $\blacksquare =$ _____

- Commutative Property of Addition

- Associative Property of Addition

- When you subtract 0 from any number, the result is the number.

Which subtraction rule is like the opposite of the Identity Property of Addition? Explain your reasoning.

Independent Practice

Algebra Find each unknown. Write each property or rule that is used.

4. (■ + 8) + 7 = 9 + (8 + 7)

■ = _____

_____

5. 14 + 13 = 13 + ■

■ = _____

_____

6. ■ + 0 = 19

■ = _____

_____

7. 25 − ■ = 0

■ = _____

_____

8. 17 + (11 + 18) = (17 + ■) + 18

■ = _____

_____

9. 37 − ■ = 37

■ = _____

_____

Use the properties of addition to add.

10. 17 + 0 = _____

11. (22 + 35) + 15 = _____

12. 16 + 22 = _____

13. 0 + 47 = _____

14. 19 + (61 + 15) = _____

15. 27 + (43 + 16) = _____

16. 23 + 74 = _____

17. (24 + 24) + 16 = _____

18. 0 + 83 = _____

19. 25 + (35 + 19) = _____

Problem Solving

20. Paco has 75 minutes before he needs to get ready for baseball practice. He cleans his room for 40 minutes and reads for 35 minutes. How much time will he have left before his baseball practice? Explain.

_____

_____

21. **Processes &Practices** **7** **Identify Structure** Chloe ate 10 grapes and 5 crackers. Layla ate 5 grapes and 10 crackers. Who ate more food items? Write a number sentence. Then identify the property or rule used.

_____

_____

Brain Builders

22. **Processes &Practices** **2** **Use Number Sense** $(23 + \blacksquare) + 19 = 23 + (\blacksquare + 19)$
Can any number complete the number sentence? Can zero be used? Explain.

_____

_____

23. **Building on the Essential Question** How are addition properties and subtraction rules helpful when solving problems? Give an example.

_____

_____

_____

MY Homework

Homework Helper

eHelp

Need help? connectED.mcgraw-hill.com

Add (44 + 18) + 22 mentally.

Use the Associative Property of Addition to make these numbers easier to add. The way in which numbers are grouped when added does not affect the sum.

(44 + 18) + 22 = 44 + (18 + 22)   ◄─ Find 18 + 22 first.

$\qquad\qquad\qquad$ = 44 + 40

$\qquad\qquad\qquad$ = 84

So, 44 + 18 + 22 = 84.

Practice

Complete each number sentence. Identify the property or rule used.

1. 85 + 0 = _____

_____

2. 96 + 13 = 13 + _____

_____

3. _____ − 0 = 37

_____

_____

4. (15 + 23) + 7 = 15 + (_____ + 7)

_____

Problem Solving

5. While bird watching, Gabrielle saw 6 robins and 3 blue jays. Chase saw 3 robins and 6 blue jays. Who saw more birds? Tell which property you used.

_____

_____

Brain Builders

6. **Processes &Practices** **Use Mental Math** For homework, Brooke has 15 math problems, some social studies problems, and 5 science problems. She has 29 problems in all. Use mental math to determine how many problems she has for social studies. Tell which property you used.

_____

7. A soccer team scored 2 goals in the first half. If they won the game by a score of 2 goals to 1 goal, how many goals did they score in the second half? Tell which property you used. Explain your thinking to a friend.

_____

_____

Vocabulary Check

Write a number sentence that demonstrates each property.

8. Commutative Property of Addition _____

9. Associative Property of Addition _____

10. Identity Property of Addition _____

11. **Test Practice** Which number sentence represents the Commutative Property of Addition?

- (A) $357 + 0 = 357$
- (B) $(7 + 19) + 3 = 7 + (19 + 3)$
- (C) $36 + 14 = 14 + 36$
- (D) $79 - 79 = 0$

Name ....

Lesson 2
Addition and Subtraction Patterns

ESSENTIAL QUESTION
What strategies can
I use to add or subtract?

 Math in My World [Tutor]

Example 1

On Friday, 1,323 people saw the new movie at a local theater. On Saturday, 1,000 more people saw the new movie. On Sunday, 100 less people saw the movie than on Saturday. How many people saw the movie each day?

 Cut 4 strips of paper. Place each strip of paper on each of the columns shown. Cover the numbers.

1,000	100	10	1
1,000	100	10	1
1,000	100	10	1

 Slide the strips of paper up to show 1 thousand, 3 hundreds, 2 tens, and 3 ones.

 Slide the paper up one row in the thousands column to show the number of people who saw the movie on Saturday.

	thousands	hundreds	tens	ones
Friday →	1	3	2	3
Saturday →	2	3	2	3
Sunday →				

 Slide the paper down one row in the hundreds column to show the number of people who saw the movie on Sunday. Write this number in the place-value chart.

So, _____ people saw the movie on Saturday and _____ saw the movie on Sunday.

Copyright © McGraw-Hill Education   Nick White/Photodisc/Getty Images

Online Content at [connectED.mcgraw-hill.com]

Lesson 2   67

Example 2

Miss Starcher drew a puzzle on the whiteboard. The puzzle contains a pattern. Solve her puzzle by filling in the two blank boxes.

 1 Each number in the first column

is _____ more than the number

in the row above it.

So, 79,245 + 1,000 = _____ .

2 Each number in the last column is

_____ more than the number

in the row above it.

So, 80,525 + 10 = _____ .

Check

Each number in the last row is _____ more than the number before it.

Since _____ + 100 = 80,345, the answer in the first column

is correct.

Guided Practice

Write each number.

1. 1,000 more than 3,872

2. 10 less than 221

_____

_____

Complete the table.

	Start	End	Change
3.	37,828	38,828	
4.	830,174		100,000 less

Complete each number sentence.

5. 36,525 + _____ = 36,625

6. 98,264 − _____ = 88,264

What do you look for when you look for a number pattern?

Name _____

Write each number.

7. 100 less than 37,972 _____

8. 10,000 more than 374 _____

9. 10 more than 45,301 _____

10. 1 more than 12,349 _____

11. 10,000 less than 12,846 _____

12. 1,000 more than 91,928 _____

13. 1 less than 37,937 _____

14. 1,000 less than 82,402 _____

Complete the table.

	Start	End	Change
15.	28,192		100 less
16.	8,392	8,402	
17.	521,457	520,457	
18.	51,183		1 more

Complete each number sentence.

19. 45,311 + _____ = 46,311

20. 28,400 − _____ = 28,390

21. 89,420 − _____ = 89,320

22. 84,552 + _____ = 94,552

23. 6,339 + _____ = 6,340

24. 3,014 + _____ = 13,014

Identify and complete each number pattern.

25. | 8,901 | 8,911 | 8,921 | | | _____ more |

26. | | 969,987 | 979,987 | | 999,987 | _____ more |

27. | 56,789 | | 56,589 | 56,489 | 56,389 |

28. | 42,578 | | | 42,608 | 42,618 |

Copyright © McGraw-Hill Education

Problem Solving

29. Go up the ladder. Write the resulting number on each rung.

	100 more
	10 more
	1,000 more
	1 more
	10,000 more
	10 more
	1,000 more
272	**Start**

30. Go down the ladder. Write the resulting number on each rung.

12,393	**Start**
	10,000 less
	100 less
	100 less
	1,000 less
	1,000 less
	100 less
	10 less

Brain Builders

31. Processes &Practices ③ **Find The Error** Gary completed this number pattern. Find and correct his mistake. Explain how you know.

27,389; 26,389; 25,389; 23,389; 24,389

_____

_____

_____

32. Processes &Practices ② **Use Number Sense** If the cost of beverages creates a pattern, what would be the cost of the quart carton of orange juice?

$0.69   [   ] $2.89 $3.99

33. Building on the Essential Question Why do we study patterns in mathematics?

_____

_____

Name

MY Homework

Homework Helper

Need help? connectED.mcgraw-hill.com

Identify and complete the number pattern.

12,345; 13,345; _____ ; 15,345; 16,345

Look to see how each number is different from the number before it.

Write the first number.

Write the second number.

1 2,345

The place that
changed in value
is the thousands.

1 3,345

The pattern is 1,000 more or + 1,000.
So, to complete the pattern, add 1,000.

$$\begin{array}{r} 13,345 \\ +\quad 1,000 \\ \hline 14,345 \end{array}$$

Check to make sure the pattern continues.

12,345   13,345   14,345   15,345   16,345

+ 1,000   + 1,000   + 1,000   + 1,000

So, the missing number is 14,345.

Practice

Write each number.

1. 100 less than 877

2. 10,000 more than 6,310

3. 10 more than 1,146

4. 1,000 less than 9,052

5. 1,000 more than 37,542

6. 10 less than 2,727

Complete each number sentence.

7. $1{,}100 + \underline{\hphantom{xxxxx}} = 1{,}200$

8. $40{,}619 - \underline{\hphantom{xxxxx}} = 39{,}619$

9. $63{,}088 - \underline{\hphantom{xxxxx}} = 53{,}088$

10. $4{,}514 + \underline{\hphantom{xxxxx}} = 4{,}524$

Complete each pattern.

11. $7{,}213; \underline{\hphantom{xxxxx}}; 7{,}413; 7{,}513$

12. $32{,}877; 42{,}877; 52{,}877; \underline{\hphantom{xxxxx}}$

13. $967, 957, \underline{\hphantom{xxxxx}}, 937$

14. $3{,}222; \underline{\hphantom{xxxxx}}; 3{,}220; 3{,}219$

Brain Builders

15. Cady had 3,469 pennies in January. In February, she had 3,539. In March, she had 3,609. If the pattern continues, how many pennies will Cady have in each month through July?

_____

_____

16. Processes &Practices **8** **Look for a Pattern** Lito keeps track of books in a warehouse. Every month he updates a chart that shows the number of books in stock. For one title, his chart shows 54,550; 44,450; 34,350. If the pattern continues, what number will Lito's chart show next month?

_____

17. Angela uses a large pitcher to water her plants. When the pitcher is full, it holds 384 ounces. Angela gives each plant 100 ounces of water. How much water is left in the pitcher after Angela waters three plants?

_____

18. Test Practice Identify the number pattern. 21,344; 20,344; 19,344

ⓐ 10 less

ⓒ 1,000 less

ⓑ 10,000 less

ⓓ 100 less

Name
....

Lesson 3
Add and Subtract Mentally

ESSENTIAL QUESTION ?
What strategies can I use to add or subtract?

To mentally add or subtract larger numbers, you can take and give to make one number end in a ten, hundred, or thousand.

 Math in My World **Tutor**

Example 1

The table shows the number of instruments sold at an instrument store. What is the total number of guitars and trumpets that was sold?

Instruments Sold	
Instrument	**Number Sold**
guitar	223
trumpet	67

Find 223 + _____.

Make a ten.

Take from one addend. → $223 + 67$ ← Give to the other addend.

$\quad\quad -3 \quad +3$

$\quad\quad 220 \quad 70$

220 + 70 = _____ ← Write a number sentence.

So, there were _____ guitars and trumpets sold.

Example 2

Find 184 − 59.

Make a ten. 59 is close to 60.

Add 1 to 59 to make 60.

$184 - 60 = 124$

Since you subtracted 1 too many, add it back in. $124 + 1 = 125$

So, 184 − 59 = _____.

Copyright © McGraw-Hill Education SW Productions/Photodisc/Getty Images

Online Content at ⟋ **connectED.mcgraw-hill.com**

Lesson 3 73

Example 3

Attendance at a concert was 82,000 people. The following week, there were 76,000 people. How many more people attended the concert the first week?

Find 82,000 − 76,000.

Both numbers have the same greatest place-value position.

Helpful Hint

You can also think

82 thousands − 76 thousands =

_____ thousands.

TTH	TH	H	T	O
8	2	0	0	0
7	6	0	0	0

The greatest place-value position is ten thousands.

1 First subtract the ten thousands and the thousands.

82 − 76 = _____

2 The difference, 6, is in the thousands place.

So, 82,000 − 76,000 = _____.

So, _____ thousand more people attended the first concert.

Guided Practice

Make a ten, hundred, or thousand to mentally add.

1.   57 + 58

− ☐ + 2
_____ _____

☐ + 60 = _____

So, 57 + 58 = _____.

2.   499 + 77

+ 1 − ☐
_____ _____

500 + ☐ = _____

So, 499 + 77 = _____.

Use mental math to subtract.

3. 184 − 76

Make a ten. 76 + 4 = _____

184 − _____ = _____

104 + 4 = _____

So, 184 − 76 = _____.

Talk MATH

Look at Exercise 3. Explain why you added 4 to the difference of 104 before writing the final answer.

Name

....

Independent Practice

4. 8,825 − 6,397

Make a hundred.

6,397 + 3 = _____

_____ − 6,400 = _____

2,425 + 3 = _____

So, 8,825 − 6,397 = _____ .

5. 684 − 169

Make a ten.

169 + 1 = _____

_____ − 170 = _____

514 + 1 = _____

So, 684 − 169 = _____ .

Make a ten, hundred, or thousand to mentally add.

6. 738 + 56 = _____

7. 223 + 728 = _____

8. 6,627 + 3,315 = _____

9. 5,478 + 1,312 = _____

Use mental math to subtract.

10. 7,930 − 4,623 = _____

11. 5,547 − 2,539 = _____

12. 8,329 − 7,218 = _____

13. 3,273 − 1,256 = _____

Subtract. Draw a line to the difference in the second column.

14. 15,000 − 8,000

• 103,000

15. 77,000 − 65,000

• 12 thousand

16. 394,000 − 44,000

• 7 thousand

17. 273,000 − 170,000

• 350,000

Copyright © McGraw-Hill Education

Problem Solving

Use mental addition or subtraction to solve.

18. Attendance at a concert was 12,769. The following evening, the attendance was 13,789. How many more people attended the concert on the second night?

19. **Processes &Practices** ▶2 **Reason** Yolanda is finding 23,567 − 12,458. She added 2 to 12,458 before subtracting. Then she added it in again after she subtracted. Is her method correct? Explain.

_____

_____

_____

Brain Builders

20. **Processes &Practices** ▶1 **Make Sense of Problems** Mentally add or subtract to get to the finish line. Explain how you know when to add or subtract.

Start	3,829	4,829		
				6,729
			16,629	
315,629		115,629	15,629	
315,639				
	415,649			715,649
				Finish

21. **Building on the Essential Question** Why is mental addition and subtraction important when you learn more difficult concepts?

_____

_____

_____

Name ....

MY Homework

Homework Helper

Need help? connectED.mcgraw-hill.com

Find 237 + 48.

Make a ten to mentally add.   237   +   48

Give to the other addend. → + 3   − 3 ← Take from one addend.

240   +   45   =   285

So, 237 + 48 = 285.

Find 752 − 23.

752 − 20 ← Change 23 to 20 by subtracting 3.

752 − 20 = 732

732 − 3 = 729 ← Since you subtracted 3 too few, subtract 3 more from the total.

So, 752 − 23 = 729.

Practice

Make a ten, hundred, or thousand to mentally add.

1. 118 + 203 = _____

2. 549 + 24 = _____

3. 1,198 + 46 = _____

4. 745 + 997 = _____

Use mental math to subtract.

5. 982 − 56 = _____

6. 7,499 − 4,100 = _____

Problem Solving

Use mental math to solve.

7. The cafeteria sells 498 cartons of milk and 246 bottles of juice every day. How many total milk and juice containers are sold each day?

8. **Processes &Practices** 5️⃣ **Use Math Tools** Terrell had a chain of 56 paper clips. Some of the paper clips fell off. Now there are 38 paper clips in the chain. How many fell off?

Brain Builders

9. Li counts 203 stars in the sky. The next night, she counts 178 stars. After three nights, she had counted a total of 500 stars. How many stars did Li see on the third night? Write a number sentence to show how you used mental math to solve.

10. **Processes &Practices** 6️⃣ **Be Precise** There were 132 children at the museum on Saturday. On Sunday, there were 61 children. How many children were there on Saturday and Sunday? Explain how to use mental math to solve.

11. **Test Practice** Mikah's dog has 39 spots. Jonelle's dog has 85 spots. How many more spots does Jonelle's dog have?

  Ⓐ 45 spots   Ⓒ 41 spots

  Ⓑ 46 spots   Ⓓ 44 spots

Name ....

Lesson 4
Estimate Sums and Differences

ESSENTIAL QUESTION ?
What strategies can I use to add or subtract?

When estimating, you can round to any place value.

 Math in My World Watch Tutor

Example 1

The Central School District needs 5,481 forks and 2,326 spoons for a school function. About how many forks and spoons will they need all together?

Estimate 5,481 + 2,326. Round to the hundreds place.

Round each number to the nearest hundred. Then add.

$$
\begin{array}{cc}
5{,}481 & \longrightarrow \boxed{\text{rounds to}} \longrightarrow \quad \boxed{}\,,\,\boxed{}\,\boxed{}\,\boxed{} \\
+\,2{,}326 & \longrightarrow \boxed{\text{rounds to}} \longrightarrow \ +\ \boxed{}\,,\,\boxed{}\,\boxed{}\,\boxed{} \\
\hline
& \boxed{}\,,\,\boxed{}\,\boxed{}\,\boxed{}
\end{array}
$$

So, 5,481 + 2,326 is about _____.

Example 2 Watch Tutor

Estimate \$7,542 − \$3,225. Round to the hundreds place.

$$
\begin{array}{cc}
\$7{,}542 & \longrightarrow \boxed{\text{rounds to}} \longrightarrow \ \$\ \boxed{}\,,\,\boxed{}\,\boxed{}\,\boxed{} \\
-\,\$3{,}225 & \longrightarrow \boxed{\text{rounds to}} \longrightarrow \ -\ \$\ \boxed{}\,,\,\boxed{}\,\boxed{}\,\boxed{} \\
\hline
& \$\ \boxed{}\,,\,\boxed{}\,\boxed{}\,\boxed{}
\end{array}
$$

So, \$7,542 − \$3,225 is about _____.

Online Content at connectED.mcgraw-hill.com

Example 3

Tutor

The table shows populations for two cities in Kentucky. About how many more people live in Covington than in Ashland?

Kentucky Populations	
City	Population
Ashland	21,510
Covington	42,811

Round each population to the nearest thousand. Then, subtract.

42,811 — rounds to → ☐☐,☐☐☐

− 21,510 — rounds to → − ☐☐,☐☐☐

‾‾‾‾‾‾‾‾‾‾‾‾‾‾‾‾‾‾‾‾

☐☐,☐☐☐

So, Covington has about _____ more people.

Guided Practice

Talk MATH

Estimate 829 + 1,560 to the nearest hundred and the nearest thousand.

Estimate. Round each number to the given place value.

1. 1,454 + 335; hundreds

_____ + _____ = _____

2. 2,871 + 427; hundreds

_____ + _____ = _____

3. $2,746 − $1,529; tens

_____ − _____ = _____

4. 48,344 − 7,263; thousands

_____ − _____ = _____

Name

Independent Practice

Estimate. Round each number to the given place value.

5. $5,238 + $3,420; hundreds

6. $4,127 + $2,666; hundreds

7. 5,342 + 298; hundreds

8. 3,182 + 6,618; hundreds

9. 48,205 + 50,214; thousands

10. $25,497 + $54,088; ten thousands

11. $7,172 − $5,103; hundreds

12. 9,185 − 6,239; thousands

13. 2,647 − 256; hundreds

14. 27,629 − 5,364; thousands

15. $27,986 − $4,521; thousands

16. $47,236 − $20,425; thousands

Copyright © McGraw-Hill Education

Problem Solving

The table shows the tallest buildings in the world. Round each height to the nearest hundred. Write a number sentence to solve.

Tallest Buildings in the World		
Building	**Location**	**Height (ft)**
Taipei 101	Taiwan	1,669
Petronas Towers	Malaysia	1,482
Willis Tower	United States	1,450
Jin Mao Building	China	1,381
CITIC Plaza	China	1,282
Shun Hing Square	China	1,259
Empire State Building	United States	1,250

17. About how much taller is the Willis Tower than the Jin Mao Building?

_____

_____

18. **Processes &Practices** 4 **Model Math** Estimate the difference between the height of the Taipei 101 building and the Empire State Building.

_____

19. About how much taller is Petronas Towers than the Empire State Building? Tell a friend how you rounded to find your answer.

_____

_____

Brain Builders

20. **Processes &Practices** 2 **Reason** Write two numbers that when rounded to the thousands place have an estimated sum of 10,000. Write a number sentence to support your answer.

_____

21. **Building on the Essential Question** How do you know if an estimate is reasonable? Explain.

_____

_____

Name ....

Homework Helper

Need help? connectED.mcgraw-hill.com

Estimate 468 + 2,319. Round to the nearest hundred.

$$
\begin{array}{r}
468 \\
+\,2{,}319 \\
\end{array}
\quad
\begin{array}{c}
\text{rounds to} \\
\text{rounds to} \\
\end{array}
\longrightarrow
\begin{array}{r}
500 \\
+\,2{,}300 \\
\hline
2{,}800 \\
\end{array}
$$

So, 468 + 2,319 is about 2,800.

Estimate 55,599 − 22,782. Round to the nearest thousand.

$$
\begin{array}{r}
55{,}599 \\
-\,22{,}782 \\
\end{array}
\quad
\begin{array}{c}
\text{rounds to} \\
\text{rounds to} \\
\end{array}
\longrightarrow
\begin{array}{r}
56{,}000 \\
-\,23{,}000 \\
\hline
33{,}000 \\
\end{array}
$$

So, 51,599 − 22,782 is about 33,000.

Practice

Estimate. Round each number to the nearest hundred.

1. 7,392 — rounds to → ☐,☐☐☐
     + 4,112 — rounds to → + ☐,☐☐☐
     ———————————————— ☐☐,☐☐☐

2. 8,752 — rounds to → ☐,☐☐☐
     − 3,269 — rounds to → − ☐,☐☐☐
     ———————————————— ☐,☐☐☐

Estimate. Round each number to the nearest thousand.

3. $5,486 + $8,602

4. 95,438 − 62,804

Problem Solving

Estimate. Round each number to the nearest hundred.

5. A total of 2,691 people attended the school play. A total of 1,521 people attended the band concert. About how many more people attended the play than the concert?

Brain Builders

Estimate. Round each number to the nearest thousand.

6. The highest point in Texas, Guadalupe Peak, is 8,749 feet high. The highest point in California, Mount Whitney, is 14,497 feet high. About how much higher is Mount Whitney than Guadalupe Peak? If you round to the nearest hundred, what is the difference between your estimates?

7. **Processes &Practices** **2** **Use Number Sense** Maria's school raised $23,240 in magazine sales and Cole's school raised $16,502. About how much more money did Maria's school raise? Explain how you rounded to estimate the difference.

8. **Test Practice** Which is the correct estimate for 63,621 − 41,589 rounded to the nearest hundred?

(A) 22,040

(C) 20,000

(B) 22,000

(D) 22,032

Check My Progress

Vocabulary Check

1. Each mouse is using a property of addition. Find each unknown. Draw lines through the maze to help each mouse find the cheese with the property that matches its addition sentence(s).

56 + 13 = _____
13 + 56 = _____

Commutative   Identity   Associative

42 + 38 = _____
38 + 42 = _____

0 + 63 = _____

(24 + 14) + 53 = _____
24 + (14 + 53) = _____

(62 + 18) + 45 = _____
62 + (18 + 45) = _____

78 + 0 = _____

Concept Check

Write each number.

2. 1,000 less than 49,737

3. 10,000 more than 53,502

Problem Solving

4. Sarah and her mom are at the mall. They buy a shirt that costs $16, a belt that costs $8, and a dress that costs $22. To find the total cost, Sarah adds $16 and $8, and then adds that sum to $22. Her mom adds $16 to the sum of $8 and $22. What property of addition are they using to find the total cost? What is the total cost?

Brain Builders

5. Mr. Cleff wants to buy the following instruments for band class.

$362 $134 $163

About how much money will he need to buy the instruments above? Write a number sentence to support your answer.

Explain how you rounded to find your estimate.

6. **Test Practice** Which number sentence can be used to estimate 3,401 + 8,342?

Ⓐ 3,000 + 8,000 = 11,000   Ⓒ 4,000 + 8,000 = 12,000

Ⓑ 3,000 + 9,000 = 12,000   Ⓓ 4,000 + 9,000 = 13,000

Lesson 5
Add Whole Numbers

ESSENTIAL QUESTION
What strategies can I use
to add or subtract?

When you add numbers, it might be necessary
to regroup.

 Math in My World

I got it!

Example 1

**There were 6,824 baseball tickets sold last week.
This week, there were 349 baseball tickets sold.
How many tickets were sold all together?**

Find 6,824 + 349.

 Add ones.

4 + 9 = 13

Regroup 13 ones as 1 ten and 3 ones.

 Add tens.

1 + 2 + 4 = 7

 Add hundreds.

8 + 3 = 11

Regroup 11 hundreds as 1 thousand and 1 hundred.

 Add thousands.

1 + 6 = 7

So, _____ tickets were sold all together.

Check for Reasonableness The estimate is _____.

Since _____ is close to the estimate, the answer is reasonable.

Example 2

Ticket sales for a play are shown in the table. What were the total sales?

Ticket Sales	
Day	Amount
Saturday	$58,713
Sunday	$43,827

Estimate

$58,713	rounds to	$60,000
+ $43,827	rounds to	+ $40,000
		$100,000

 Add ones.

3 + 7 = 10

Regroup 10 ones as 1 ten and 0 ones.

1 + 1 + 2 = 4

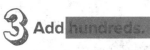

7 + 8 = 15

Regroup 15 hundreds as 1 thousand and 5 hundreds.

 Add thousands.

1 + 8 + 3 = 12

Regroup 12 thousands as 1 ten thousand and 2 thousands.

1 + 5 + 4 = 10

Place the $ symbol in front of the sum.

So, the total ticket sales were _____.

Explain why it is important to line up digits in numbers when you add.

Guided Practice

Add. Estimate to check your work.

1.	$2,961	2.	29,380
	+ $4,205		+ 10,225

Independent Practice

Add. Estimate to check your work.

3. 8,346
+ 7,208

4. $23,824
+ $7,346

5. 82,828
+ 4,789

6. $37,178
+ $82,370

7. $693,782
+ $47,816

8. 743,980
+ 211,315

9. 254,671
+ 381,366

10. $15,789
+ $22,503

11. 56,772
+ 29,428

Add. Use the place-value chart to help set up the problem.

12. 17,599 + 72,682 = _____

	Thousands			Ones		
	hundreds	tens	ones	hundreds	tens	ones
+						

Problem Solving

13. There are 4,585 students who rode the bus to school today. There were 3,369 students who came to school another way. How many students were there in all at the school?

_____

Brain Builders

14. **Processes &Practices** **6** **Explain to a Friend** Becky's mom wants to buy a new TV that costs $1,579 and a DVD player that costs $247. She has $2,000. If she buys groceries for $136, will she have enough money for the TV and the DVD player? Explain to a friend.

_____

_____

15. Mr. Russo's class is collecting bottles to recycle. The class collected some bottles in March and 2,555 bottles in April. 3,701 bottles were collected in all. How many were collected in March?

_____

16. **Processes &Practices** **1** **Make Sense of Problems** Write two 5-digit addends whose sum would give an estimate of 60,000. Explain your reasoning.

_____

_____

17. **Building on the Essential Question** Explain why an addition problem that has 4-digit addends could have a 5-digit sum.

_____

_____

Name ....

MY Homework

Homework Helper

Need help? ↗ connectED.mcgraw-hill.com

Find 32,866 + 7,375.

Estimate 32,866   rounds to   33,000
          + 7,375   rounds to   + 7,000
                                  40,000

1 Add ones.
6 + 5 = 11
Regroup 11 ones as 1 ten and 1 one.

2 Add tens.
1 + 6 + 7 = 14
Regroup 14 tens as 1 hundred and 4 tens.

3 Add hundreds.
1 + 8 + 3 = 12
Regroup 12 hundreds as 1 thousand and 2 hundreds.

4 Add thousands.
1 + 2 + 7 = 10
Regroup 10 thousands as 1 ten thousand and 0 thousands.

5 Add ten thousands.
1 + 3 = 4

So, 32,866 + 7,375 = 40,241.

40,241 is close to the estimate of 40,000. The answer is reasonable.

Practice

Add. Estimate to check your work.

1.   5,239
        + 2,794

2. $4,189
        + $5,432

3.   169,748
        + 355,470

4.   452,903
        + 318,766

Brain Builders

5. **Processes &Practices** **5** **Use Math Tools** A zoo has two elephants, Sally and Joe. Sally weighs 7,645 pounds, and Joe weighs 12,479 pounds. The maximum weight to cross a bridge is 20,000 pounds. Should Sally and Joe cross the bridge together? Explain.

_____

_____

6. At a library, 1,324 children's books and 1,510 fiction books were checked out. How many books were checked out at the library? Explain how you know to a friend.

_____

_____

7. **Test Practice** Find the unknown in $45,209 + $31,854 = ▪.

Ⓐ $76,063

Ⓒ $77,053

Ⓑ $77,163

Ⓓ $77,063

92 **Need more practice?** Download Extra Practice at connectED.mcgraw-hill.com

Copyright © McGraw-Hill Education

Name
....

Subtraction of whole numbers is similar to addition of whole numbers because you might need to regroup.

 Math in My World | Watch | Tools | Tutor

Example 1

The Trevino family is moving to a new city. They have driven 957 miles out of the 3,214 miles that they need to drive. How many more miles do they need to drive?

Find $3,214 - 957$.

 Subtract ones.
Regroup a ten as 10 ones.
10 ones + 4 ones = 14 ones

14 ones − 7 ones = _____ ones

 Subtract tens.
Regroup a hundred as _____ tens.
10 tens + 0 tens = 10 tens

10 tens − 5 tens = _____ tens

 Subtract hundreds.
Regroup a thousand as _____ hundreds.
10 hundreds + 1 hundred = 11 hundreds

11 hundreds − 9 hundreds = _____ hundreds

$$\begin{array}{r} 3,\ 2\ \ 1\ \ 4 \\ -\quad 9\ \ 5\ \ 7 \\ \hline \end{array}$$

 Subtract thousands.
2 thousands − 0 thousands = _____ thousands

So, $3,214 - 957 =$ _____. The Trevino family

needs to drive _____ more miles.

Online Content at 🖱 **connectED.mcgraw-hill.com**

The **minuend** is the first number in a subtraction sentence from which a second number is to be subtracted. The **subtrahend** is the number that is subtracted.

Example 2

The band has raised $1,345 toward new equipment. If the goal is to raise $4,275, how much money must still be raised?

Estimate $4,275   rounds to → $4,300
              −$1,345   rounds to → $1,300
                                   $3,000

 Subtract ones.
5 − 5 = 0

 Subtract tens.
7 − 4 = 3

 Subtract hundreds.
Regroup a thousand as 10 hundreds.
12 − 3 = 9

Subtract thousands.
3 − 1 = 2

So, the band still needs to raise _____.

Check for Reasonableness You can use addition to check your subtraction.

```
  4,275          2,930
−  1,345       + 1,345
  2,930          4,275
```

The answer is correct and close to the estimate.

Explain how to check the answer to a subtraction problem by using addition.

Guided Practice

Subtract. Use addition or estimation to check.

1. 2,962   2. $4,785
  −   845   − $2,293

Independent Practice

Subtract. Use addition or estimation to check.

3.
```
   8,845
 −   627
```

4.
```
   $5,751
 − $4,824
```

5.
```
   $8,327
 − $5,709
```

6.
```
   39,536
 − 18,698
```

7.
```
   847,311
 − 562,530
```

8.
```
   93,458
 − 21,649
```

9.
```
   78,215
 − 56,827
```

10.
```
   $18,345
 − $14,400
```

11.
```
   629,843
 − 216,954
```

Subtract. Use addition or estimation to check. Use the place-value chart to set up the problem.

12. 961,344 − 345,822 = _____

Thousands			Ones		
hundreds	tens	ones	hundreds	tens	ones

13. Do you prefer to use addition or estimation to check? Explain.

_____

_____

Brain Builders

14. **Processes & Practices** **Use Math Tools** There are a total of 1,569 tickets for a concert. On the first day of sales, 875 tickets were sold. The following day an additional 213 tickets were sold. How many tickets are still available?

15. Mount Everest is 29,029 feet high. From base camp at 17,600 feet, a climber hiked 2,300 feet. How much farther does the climber have before reaching the summit of the mountain?

16. John Adams was born in 1732 and became President in 1797. Harry S. Truman was born in 1884 and became President in 1945. Who was older when he became President? How much older?

17. **Processes & Practices** **Use Number Sense** Circle the subtraction problem that does not require regrouping. Explain.

| 67,457 | 71,639 | 89,584 | 95,947 |
| − 40,724 | − 39,607 | − 57,372 | − 26,377 |

18. **Building on the Essential Question** Why is it important to line up the digits in each place-value position when subtracting?

MY Homework

Homework Helper

Need help? connectED.mcgraw-hill.com

Find 6,325 − 2,841.

Estimate Round to the nearest thousand. 6,000 − 3,000 = 3,000

1 Subtract ones.

2 Subtract tens.
Regroup a hundred as 10 tens.

3 Subtract hundreds.
Regroup a thousand as ten hundreds.

4 Subtract thousands.

```
      5    12   12
     6 ,  3    2   5
  −  2 ,  8    4   1
     3 ,  4    8   4
```

So, 6,325 − 2,841 = 3,484.

Check

Use addition to check.

```
  6,325          3,484
− 2,841        + 2,841
  3,484          6,325
```

So, the answer is reasonable.

Practice

Subtract. Use addition or estimation to check.

1. $6,148
    − $1,575

2. 9,516
    − 7,228

3. 6,637
    − 2,846

4. 33,539
    − 31,649

Problem Solving

5. A minor league baseball team gave out 1,250 free hats. If 2,359 people attended the game, how many did not get a hat?

Brain Builders

6. **Processes &Practices** **Use Math Tools** There were 1,400 shirts and 2,115 shirts at the two stadium stores before the game. After the game, there were 1,396 shirts left. How many were sold during the game? Write a number sentence to support your answer.

Vocabulary Check [Vocab $^a{}_b{}_c$]

7. Label each part of the subtraction problem with the correct vocabulary term.

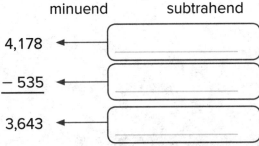

difference   minuend   subtrahend

$$
\begin{array}{r}
4{,}178 \\
-\ 535 \\
\hline
3{,}643
\end{array}
$$

8. **Test Practice** Find the unknown in $1{,}515 - 1{,}370 = \blacksquare$.

    Ⓐ 165   Ⓒ 135

    Ⓑ 145   Ⓓ 235

Name _____

ESSENTIAL QUESTION
What strategies can I use to add or subtract?

Sometimes subtraction involves minuends that have zeros.

 Math in My World Watch Tutor

Example 1

Each fourth grade class has a goal to collect 5,100 pennies for charity. How many more pennies do the fourth graders in Mr. Blake's class need to reach their goal?

Find 5,100 − 3,520.

Class	Pennies
Mrs. Clark	4,523
Mr. Blake	3,520
Ms. Simms	1,987
Mrs. Stone	2,569

 Subtract ones.
0 ones − 0 ones = 0 ones

 Subtract tens.
Regroup 1 hundred as 10 tens.

10 tens − 2 tens = 8 tens

 Subtract hundreds.
Regroup one thousand as 10 hundreds.

10 hundreds − 5 hundreds = 5 hundreds

 Subtract thousands.
4 thousands − 3 thousands = 1 thousand

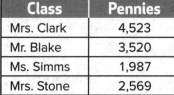

So, Mr. Blake's class needs _____ more pennies.

Check

```
    □ , □ □ □      The answer is correct.
  + 3 , 5  2  0
    5 , 1  0  0
```

Example 2

There were 30,090 fans at the stadium on Saturday. The next Saturday, there were 22,977 fans. How many more fans were at the stadium on the first Saturday than the second Saturday?

Find 30,090 − 22,977.

 Subtract ones.

Regroup 1 ten as 10 ones.

10 ones − 7 ones = _____ ones

 Subtract tens.

8 tens − 7 tens = _____ ten

3 Subtract hundreds.

Regroup 1 ten thousand as 10 thousands.
Regroup 1 thousand as 10 hundreds.

10 hundreds − 9 hundreds = _____ hundred

4 Subtract thousands.

9 thousands − 2 thousands = _____ thousands

 Subtract ten thousands.

2 ten thousands − 2 ten thousands = _____ ten thousands

So, 30,090 − 22,977 = _____.

There were _____ more fans than the first Saturday.

Talk MATH

Explain how to subtract 42,956 from 55,000.

Guided Practice

Subtract. Use addition or estimation to check.

1.
```
  2, 0  0  3
- 1, 1  5  4
  ▢  ▢  ▢
```

2.
```
 $8, 0  0  0
-$3, 5  0  2
$▢ , ▢  ▢  ▢
```

Name _____

Subtract. Use addition or estimation to check.

3. 2,040
    −   946

4. 7,008
    − 2,055

5. 12,050
    − 3,162

6. 10,400
    − 5,445

7. 46,801
    − 5,823

8. 60,032
    − 21,833

9. $52,006
    − $13,055

10. 600,000
    − 28,005

11. 508,200
    − 136,118

Subtract. Use addition or estimation to check. Use the place-value chart to set up the problem.

12. 900,000 − 31,650 = _____

Thousands			Ones		
hundreds	tens	ones	hundreds	tens	ones

Problem Solving

For Exercises 13 and 14, use the table which shows the distance between New York City and five other cities around the world.

City	Miles
Jakarta, Indonesia	10,053
London, England	3,471
Mexico City, Mexico	2,086
Munich, Germany	4,042
Paris, France	3,635

13. How many more miles is it to travel to Jakarta than to London?

_____

14. How many more miles is it to travel to Munich than to Paris?

_____

Brain Builders

15. **Processes &Practices** **5** **Use Math Tools** Trent earned 3,073 and 1,932 points in two rounds of a video game. He earned 2,375 fewer points than his brother. How many points did his brother earn? Write a number sentence to support your reasoning.

_____

16. **Processes &Practices** **1** **Plan Your Solution** Identify a number that results in a 4-digit number when 156,350 is subtracted from it. Explain how to use addition to check your answer.

_____

_____

17. **Building on the Essential Question** How does understanding place value help you to subtract across zeros?

_____

_____

_____

Name ....

MY Homework

Lesson 7

Subtract Across Zeros

Homework Helper eHelp

Need help? connectED.mcgraw-hill.com

Find 10,200 − 4,795.

1 Subtract ones.
   Regroup a hundred as 10 tens.
   Regroup a ten as 10 ones.

2 Subtract tens.

3 Subtract hundreds.
   Regroup a ten thousand as 10 thousands.
   Regroup a thousand as 10 hundreds.

4 Subtract thousands.

5 Subtract ten thousands.
   0 ten thousands − 0 ten thousands = 0 ten thousands.

So, 10,200 − 4,795 is 5,405.

Practice

Subtract. Use addition or estimation to check.

1. 4,000
   − 1,731

2. 3,300
   − 1,892

3. 8,000
   − 6,313

4. $14,000
   − $10,892

Copyright © McGraw-Hill Education

Problem Solving

5. If 700 tickets were sold for a concert and only 587 people attended, how many people bought a ticket but did not attend?

6. The Amazon River, in South America, is 4,000 miles long. The Snake River, in the northwestern United States, is 1,038 miles long. How much longer is the Amazon River than the Snake River?

Brain Builders

7. **Processes &Practices** **5** **Use Math Tools** There are 6,000 products at the store. In the first hour, 327 products are sold. In the second hour, 98 products are sold. How many products are left? What was your first step to solve this problem?

8. A field of corn has 2,000 insects. There are 314 grasshoppers and 183 beetles eating the corn. How many insects are not eating the corn? Explain how you know.

9. **Test Practice** Logan has a gift card for $200. He spends $45 on Monday and $61 on Tuesday. How much money is left on his gift card?

Ⓐ $94

Ⓒ $104

Ⓑ $106

Ⓓ $139

Copyright © McGraw-Hill Education

Check My Progress

Vocabulary Check

1. Help each butterfly find its flower by drawing lines to match each vocabulary word with its definition.

minuend

subtrahend

difference

The first number in a subtraction sentence from which a second number is to be subtracted.

The answer to a subtraction problem.

The number that is subtracted in a subtraction problem.

Concept Check

Add. Estimate to check your work.

2. $3,618
    + $2,956

3. 36,847
    + 14,268

4. 529,318
    + 231,937

Subtract. Use addition or estimation to check.

5. 5,428
    − 725

6. $90,000
    − $24,074

7. 836,422
    − 145,742

Problem Solving

8. Jabar traveled 3,052 miles last year. His brother traveled 5,294 miles. How many miles did they travel in all?

9. There were 15,292 concert tickets sold last year. There were 26,935 concert tickets sold this year. How many more tickets were sold this year than last year?

Brain Builders

10. A store wants to earn $100,000 this year. So far, they have earned $42,002 and 40,050. How much more money will they have to earn to meet their goal? Explain.

11. A book has 31,225 words. Two shorter books have 14,632 words and 10,261 words. How many more words does the longer book have than the two shorter books? Explain.

12. Test Practice Find the unknown.

$24,378 + 12,489 = p$

Ⓐ $p = 36,867$ Ⓒ $p = 12,111$

Ⓑ $p = 36,757$ Ⓓ $p = 11,889$

Name ....

Lesson 8
Problem-Solving Investigation
STRATEGY: Draw a Diagram

ESSENTIAL QUESTION
What strategies can
I use to add or subtract?

Learn the Strategy Tutor

Keith's summer camp is going to build some tree houses. They will need $2,492 for tools, and $12,607 for wood. How much money do they need to build the tree houses?

1 Understand

What facts do I know?

Tools cost $_____. Wood costs $_____.

What do I need to find?

Find how much money is needed to build the tree houses.

2 Plan

I can draw a bar diagram and add to find the sum.

3 Solve

The diagram shows each part that is needed. Add to find the total.

$$
\begin{array}{r}
\$\ 2{,}4\ 9\ 2 \\
+\ \$1\ 2{,}6\ 0\ 7 \\
\hline
\$\square\square{,}\square\square\square
\end{array}
$$

So, a total of _____ is needed to build the tree houses.

4 Check

Does your answer make sense? Explain.

_____

_____

East School District has 52,672 students. West School District has 34,089 students. How many more students are there in East School District than in West School District?

 Understand

What facts do I know?

_____

_____

_____

What do I need to find?

_____

_____

Plan

_____

Solve

Check

Does your answer make sense? Explain.

_____

Apply the Strategy

Solve each problem by drawing a diagram.

1. **Processes &Practices** ▶4 **Model Math** Twickenham Stadium, in England, can seat 82,000 people. If there are 49,837 people seated in the stadium, how many more people can be seated in the stadium?

_____

Brain Builders

2. A bakery uses ten cups of butter and ten eggs for a recipe. There are 16,280 Calories in ten cups of butter. Ten eggs have 1,170 Calories. How many more Calories are there in ten cups of butter than in ten eggs?

_____

Explain how to use your diagram to check your answer.

_____

_____

_____

3. Write and solve a real-world problem that has a sum of 11,982.

_____

_____

_____

_____

Review the Strategies

Use any strategy to solve each problem.

- Use the four - step plan.
- Draw a diagram

4. Miss Bintel wants to buy a car that costs $35,500. There is a sale on Sunday. If she buys the car on Sunday, she will save $2,499. How much will the car cost on Sunday?

The car will cost _____ on Sunday.

Processes &Practices 5 **Use Math Tools** Rick drove 12,363 miles in his new car the first year he owned it. He drove 15,394 miles the second year. How many miles did Rick drive these two years?

6. Mrs. Walker has 2,005 recipes to organize. She has organized 962 of them. How many more recipes does she need to organize?

7. A moose weighs 1,820 pounds. A camel weighs 1,521 pounds. How much more does a moose weigh than a camel?

Processes &Practices 3 **Find the Error** Macy wants to find the sum of 61,043 and 23,948. Her answer is 37,095. Find and correct her mistake.

Name

Homework Helper

Need help? connectED.mcgraw-hill.com

On the first day of an audition, 2,731 singers participated. On the second day of an audition, 4,327 singers participated. How many singers participated in all? Use a bar diagram to solve.

1 Understand

What facts do I know?

There were 2,731 singers on the first day.

There were 4,327 singers on the second day.

What do I need to find?

Find how many singers there were in all.

2 Plan

I can draw a bar diagram and add to find the sum.

3 Solve

The diagram shows each part that is needed. Add to find the total.

?	
2,731	4,327

$$\begin{array}{r} 2{,}731 \\ +\ 4{,}327 \\ \hline 7{,}058 \end{array}$$

So, a total of 7,058 singers participated.

4 Check

Does your answer make sense? Explain.

2,731 rounds to 3,000. 4,327 rounds to 4,000. 3,000 + 4,000 = 7,000.
7,000 is close to the actual sum of 7,058. So, my answer makes sense.

Problem Solving

Solve each problem by drawing a diagram.

1. Joseph has 3,124 pieces of paper in his classroom. Emily has 5,229 pieces of paper in her classroom. How many pieces of paper do they have in both classrooms?

Brain Builders

2. Brayden sold 2,306 tickets for a school raffle. Connor sold 1,523 fewer tickets for a school raffle. How many tickets did Connor sell?

3. On Saturday, 5,395 people visited a museum. On Sunday, 3,118 people visited a museum. How many people visited all together on Saturday and Sunday? Explain how to use your diagram to solve this problem.

Name ....

Lesson 9
Solve Multi-Step Word Problems

ESSENTIAL QUESTION ❓
What strategies can I use
to add or subtract?

You can write an **equation** to help organize and solve
multi-step problems. An equation is a sentence that
contains an equals sign (=), showing that both sides
of the equals sign are equal.

 Math in My World Tutor

Example 1

The music club had **$390** in their account. At the
concert, they earned **$472**. Afterwards, they had to
pay **$75** to rent the stage and **$102** for the rental
equipment. How much is in their account now?

Write an equation.

| starting amount | earned at concert | paid for stage | paid for equipment | final amount |

$390 + $472 − $75 − $102 = ▪ ← unknown

When more than one step is involved, add and subtract in order from
left to right.

1 Add.

$390
+ $472
_____

2 Subtract.

_____
− $75
_____

3 Subtract.

_____
− $102
_____

So, the account now has _____ .

Check An estimate is $400 + $500 − $100 − $100, or _____ .

This is close to the actual amount, which is $ _____ .
So, the answer is reasonable.

A **variable** is a symbol, often a letter, that is used to represent an unknown, or a number that is not known.

Example 2

Twenty people got on the bus at the first stop. At the second stop, 14 people got off the bus, and 5 people got on the bus. At the third stop, 2 people got off the bus, and some people got on the bus. Then there were 24 people on the bus. How many people got on the bus at the third stop?

Write an equation.
The letter b can be used as a variable to represent the unknown.

1st stop: get on	2nd stop: get off	2nd stop: get on	3rd stop: get off	3rd stop: get on	final amount

$$20 - 14 + 5 - 2 + b = 24$$

Find the unknown.

1 Subtract.

$$\begin{array}{r} 20 \\ -14 \\ \hline \square \end{array}$$

2 Add.

$$\begin{array}{r} \square \\ +\ 5 \\ \hline \square \end{array}$$

3 Subtract.

$$\begin{array}{r} \square \\ -\ 2 \\ \hline \square \end{array}$$

4 Find the variable.

$$\begin{array}{r} 9 \\ +\ b \\ \hline 24 \end{array}$$

Since $9 + b = 24$, then $24 - 9 = b$.

$24 - 9 = \boxed{}$

$b = \boxed{}$

So, $\boxed{}$ people got on the bus at the third stop.

Guided Practice

1. **Algebra** Savannah had $15. She earned $20. Then, she bought a gift for $8. How much money does she have left? Write an equation to solve the problem. Use a variable for the unknown.

_____

_____

Can you use any letter of the alphabet for a variable? Explain.

Independent Practice

Algebra **Write an equation to solve each problem.**
Use a variable for the unknown.

2. Bailey had 75 beads. She used 20 of them on a necklace and 12 of them on a bracelet. Then she bought 25 more beads. How many beads does Bailey have now?

3. Alex had $30. He spent $13 on a game and $5 on a poster. Then, he earned $8 doing chores for a week. How much money does Alex have now?

4. Hunter had 16 jars of paint. He used 2 of them on a painting. He bought 8 more jars. Then, he used some of the jars to make another painting. Now, Hunter has 15 jars of paint. How many jars did he use for the second painting?

5. A restaurant served food to a large party. The manager is listing the total costs, shown below.

Item	Price ($)
chicken	452
pasta	388
salad	150
side dishes	s

The total cost is $1,317. How much did the side dishes cost?

Problem Solving

Use a number cube to complete each number puzzle.

6. Roll a number cube 4 times. Write one number in each box. Find the greatest value of the variable.

$\boxed{} + \boxed{} - \boxed{} + \boxed{} = b$

$b = \underline{\hspace{3cm}}$

7. Roll a number cube 6 times. Write one number in each box. Find the greatest value of the variable.

$\boxed{} + \boxed{} - \boxed{} + \boxed{} + \boxed{} - \boxed{} = y$

$y = \underline{\hspace{3cm}}$

Brain Builders

8. **Processes &Practices** **1** **Make Sense of Problems** Victoria had some money in her wallet. She went to the mall and spent $8 on a stuffed animal, $7 on lunch, and $13 on a gift for her mom. Then her brother gave her $10. She bought a book for $15. Now she has $12. How much money did Victoria originally have in her wallet? Explain your answer to a friend.

_____

_____

9. **Building on the Essential Question** How can I use variables to describe real-world problems? Explain.

_____

_____

_____

Name _____

Homework Helper

Need help? connectED.mcgraw-hill.com

The concession stand employees started with **$520** in the cash register. They earned **$725** at the football game. They had to pay **$125** for more popcorn and **$65** for more hot chocolate. How much is in the cash register now?

Write an equation.

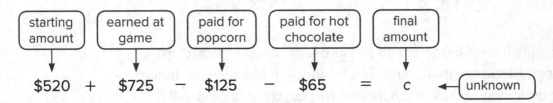

| starting amount | earned at game | paid for popcorn | paid for hot chocolate | final amount |

$520 + $725 − $125 − $65 = c ← unknown

Add and subtract in order from left to right.

Estimate 520 + 725 − 125 − 65 = c

rounds to

500 + 700 − 100 − 100 = $1,000

1 Add.

```
   520
 + 725
 1,245
```

2 Subtract.

```
 1,245
 − 125
 1,120
```

3 Subtract.

```
 1,120
 −  65
 1,055
```

So, the cash register now has $1,055.

Check The estimate is $1,000. This is close to the actual amount, which is $1,055. So, the answer is reasonable.

Problem Solving

Processes & Practices **Use Algebra** Write an equation to solve each problem. Use a variable for the unknown.

1. Trent's mom gave him $30. He earned another $12 completing chores. Trent spent $15 at the movie theater and $6 on lunch. How much money does Trent have?

_____

2. The cafeteria ordered 400 paper plates. They used 226 at breakfast. They bought 100 more. Then, they used some plates for lunch. Now there are 78 plates. How many plates did they use at lunch?

_____

Brain Builders

3. Mia's parents have $150 to spend at the beach. It will cost $70 to rent a boat, $15 for snorkels, and $25 for lunch. If Mia pays for lunch with her own money, how much money will her parents have left?

_____

Vocabulary Check

Complete each sentence using the words below.

equation   variable

4. A(n) _____ is a symbol, usually a letter, that is used to represent an unknown, or an amount that has not been identified.

5. A(n) _____ is a sentence that contains an equals sign (=), showing that both sides of the equals sign are equal.

6. **Test Practice** There are 367 boxes of graham crackers at the grocery store. On Monday, 126 boxes are sold and on Tuesday, 92 boxes are sold. On Wednesday, 203 more boxes are delivered. How many boxes are there now? Which equation represents this situation?

Ⓐ $367 + 126 + 92 - 203 = b$ Ⓑ $367 - 126 - 92 + 203 = b$

Ⓒ $367 + 126 - 92 + 203 = b$ Ⓓ $367 + 126 - 92 + b = 203$

Name ....

Add.

1. 53,035
   + 39,952

2. 94,225
   + 63,236

3. 82,427
   + 37,174

4. 32,472
   + 18,009

5. 72,259
   + 62,905

6. 52,372
   + 17,429

7. 63,141
   + 14,603

8. 20,407
   + 38,692

9. 367,028
   + 52,842

10. 482,952
    + 20,485

11. 137,953
    + 84,037

12. 813,448
    + 92,734

13. 109,374
    + 824,849

14. 372,555
    + 372,555

15. 218,662
    + 741,852

16. 359,751
    + 486,258

17. 118,577
    + 254,009

18. 888,888
    + 102,222

19. 328,805
    + 646,464

20. 335,533
    + 254,009

Name

Fluency Practice

Subtract.

1. 63,581
    − 37,510

2. 72,510
    − 62,507

3. 82,404
    − 15,840

4. 43,524
    − 43,509

5. 42,824
    − 29,131

6. 34,108
    − 19,888

7. 13,546
    − 12,816

8. 45,850
    − 29,544

9. 237,482
    − 52,851

10. 321,123
    − 32,123

11. 137,953
    − 84,037

12. 338,200
    − 12,658

13. 825,385
    − 703,261

14. 651,851
    − 215,992

15. 453,166
    − 405,556

16. 212,894
    − 198,284

17. 489,255
    − 281,816

18. 258,914
    − 168,876

19. 545,248
    − 359,249

20. 605,060
    − 488,777

Review

Vocabulary Check

Use the vocabulary words in the word bank to fill in the blanks.

Associative Property of Addition   **Commutative Property of Addition**

equation   **Identity Property of Addition**

minuend   **subtrahend**

unknown   **variable**

1. The _____ states that for any number, zero plus that number is the number.

2. A(n) _____ quantity is an amount whose value needs to be found.

3. The _____ states that the order in which two numbers are added does not change the sum.

4. The first number in a subtraction sentence from which a second number is to be subtracted is the _____.

5. The _____ states that the grouping of the addends does not change the sum.

6. A number that is subtracted from another number is called the _____.

7. A(n) _____ is a symbol, usually a letter, that is used to represent an unknown quantity.

8. A(n) _____ is a sentence that contains an equals sign (=), showing that two expressions are equal.

Concept Check

Find each unknown. Write the addition property or subtraction rule that each shows.

9. $35 - \blacksquare = 35$

10. $(16 + 5) + \blacksquare = 16 + (5 + 10)$

_____

_____

_____

_____

_____

_____

11. $83 + 35 = 35 + \blacksquare$

12. $76 + 0 = \blacksquare$

_____

_____

Write each number.

13. 10,000 more than 25,953

14. 1,000 less than 63,035

_____

_____

Make a ten, hundred, or thousand to mentally add.

15. $4,529 + 56 = $ _____

16. $506 + 349 = $ _____

Add. Estimate to check your work.

17.	**18.**	**19.**
82,267	432,901	206,522
+ 21,037	+ 177,235	+ 321,877

Subtract. Use addition or estimation to check.

20.	**21.**	**22.**
$54,751	9,004	70,909
− $43,226	− 632	− 63,485

23. Mrs. VanHorn will publish 1,045 recipes. She has published 632 of them. How many more recipes does she need to publish?

Brain Builders

24. Parker had $32. He earned $10. Then, he bought a video game for $28. He also received $10 for mowing the lawn. How much money does he have left? Write an equation using a variable for the unknown.

25. Abby won 57 points on her first turn in a game. She won 37 more points on her second turn and then lost 19 points on her third turn. She won some more points on her fourth turn. Now Abby has 100 points. How many points did she win on her fourth turn? Write an equation using a variable for the unknown. Explain your thinking to a friend.

26. Test Practice Rick drove 12,363 miles in his new car the first year he owned it. He drove 15,934 miles the second year. How many miles all together did Rick drive these two years?

Ⓐ 23,571 miles   Ⓒ 28,291 miles

Ⓑ 27,297 miles   Ⓓ 28,297 miles

Reflect

Use what you learned about addition
and subtraction to complete the grapic
organizer.

**Addition
Example**

**Subtraction
Example**

**ESSENTIAL
QUESTION**
**What strategies can I
use to add or subtract?**

Vocabulary

Estimate

Reflect on the ESSENTIAL QUESTION Write your answer below.

_____

_____

Performance Task

Brain Builders

The Ski Club

The ski club sponsors field trips and ski competitions during the school year. The club conducts fundraising activities to raise money to pay for the events.

Show all your work to receive full credit.

Part A

Eight students and one adult plan to go to Snowflake Ski Resort for a day of skiing. Student lift tickets cost $8 each and adult lift tickets cost $12 each. Lunch at the lodge costs $7 per person. Each person gets a lunch at the lodge. The resort also offers a group package for 12 people of any age that includes lift tickets and lunch for a total of $125. The club wants to spend the least amount of money possible. Should it purchase individual lunches and lift tickets or the group package? Explain your reasoning.

Part B

Club members will participate in a ski competition in February. The school made 6 buses available to transport students and families to the competition to watch the club members compete. The buses can transport 264 people to the competition. So far, 81 students, 54 parents, 27 friends, and 14 coaches have signed up to ride the bus. What is the greatest number of people that can still sign up to ride a bus to the competition? Explain your reasoning.

Part C

After a fundraising bake sale, the ski club treasurer must deduct the costs of the bake sale from the money taken in. She writes a subtraction problem in the financial book to calculate the new balance. When the financial book accidentally gets wet, two digits in the problem can no longer be read. The subtraction problem now looks like this:

$$
\begin{array}{r}
3\ 1\ 0\ 7 \\
-\ \ \ 8\ \square\ 9 \\
\hline
2\ \square\ 4\ 8
\end{array}
$$

What are the missing digits? Use your knowledge of place value to explain your reasoning.

Part D

A local organization will donate the same amount of money to the club's trip fund each week in December. The club started December with a certain amount of money already in its account. Fill out the table with the missing dollar amounts. How much did the local organization donate each week? Explain your reasoning.

End of Week	Amount in Account ($)
0	
1	363
2	488
3	613
4	

Part E

The ski club plans a year-end trip to Deep Valley Ski Resort to ski and relax. At the end of February, the club had $694 in its account. The trip costs $4,285 for meals and lodging, $2,318 for transportation, and $800 for entertainment. A donor has offered to pay $3,500 of the costs. Write an equation that can be used to find the remaining amount of money that the club must raise. Use n to represent the remaining amount. Then solve the equation. Explain your reasoning.

Copyright © McGraw-Hill Education

124PT2 **Chapter 2** Add and Subtract Whole Numbers

3 Understand Multiplication and Division

MY World of Fun

Copyright © McGraw-Hill Education   (l)©JGI/Jamie Grill/Blend Images LLC; (r)©LWA/Dann Tardif/Blend Images

ESSENTIAL QUESTION

How are multiplication and division related?

Watch

Watch a video!

MY Chapter Project

Multiples Poster

1. Now that your class has learned about multiples, it's time to illustrate some! In your group, select a number from 2 to 9 and write that number below.

Our number: _____

2. Identify 3 multiples for your number. Write these multiples on the line below.

Multiples for our number: _____

3. Your teacher will give you a sheet of poster board. Write your number at the very top. Then divide the poster into three parts, and write one of your multiples in each part.

4. Under each multiple, draw groups of items to represent the multiple. For example, if your number was 4 and your multiple was 20, draw 5 boxes of crayons with 4 crayons inside each box. This illustrates why 20 is a multiple of 4. Use the space below to brainstorm ideas with your group.

Name

....

Am I Ready?

Complete each number sentence.

1. $4 + 4 + 4 =$ _____

2. $6 + 6 +$ _____ $+ 6 = 24$

3. $7 + 7 + 7 = 3 \times$ _____

4. $9 + 9 + 9 + 9 =$ _____ $\times 9$

5. Write the multiplication fact modeled by the array to the right.

_____

Circle equal groups of 3 in each array.

6.

★ ★ ★ ★
★ ★ ★ ★
★ ★ ★ ★

7.

ⲤⲤⲤ   ⲤⲤⲤ
ⲤⲤⲤ   ⲤⲤⲤ
ⲤⲤⲤ   ⲤⲤⲤ
ⲤⲤⲤ   ⲤⲤⲤ

Complete each number pattern.

8. 2, 4, 6, _____, 10, _____, 14

9. 4, 8, 12, _____, 20, 24, _____

10. 5, _____, 15, 20, _____, 30, _____

11. _____, 18, 27, _____, 45, 54, _____

Shade the boxes to show the problems you answered correctly.

How Did I Do?

1	2	3	4	5	6	7	8	9	10	11

Name

....

MY Math Words

Vocab
$^a{}_b{}_c$

Review Vocabulary

divide   multiply

Making Connections

Use the review vocabulary to tell which operation(s) to use in each
bubble. Solve the word problems.

How many squares are
in the array all together?

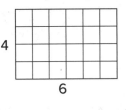

4

6

Five friends arranged
10 total baseball cards
into 5 equal groups.
How many cards are
in each group?

Divide, Multiply,

or Both?

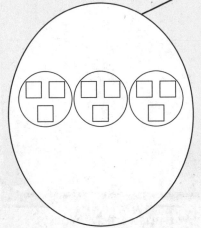

There are 5 baskets.
Each basket has
4 pieces of fruit. How
many total pieces of fruit
are there?

MY Vocabulary Cards

Vocab
abc

Processes
&Practices

Lesson 3–6

Associative Property of Multiplication

$3 \times (4 \times 6) = (3 \times 4) \times 6$

Lesson 3–5

Commutative Property of Multiplication

$3 \times 6 = 6 \times 3$

Lesson 3–7

decompose

6

$6 \times 1 = 6$
$3 \times 2 = 6$
$2 \times 3 = 6$
$1 \times 6 = 6$

Factors: 1, 2, 3, 6

Lesson 3–1

dividend

$64 \div 8 = 8$

Lesson 3–1

divisor

$3\overline{)19}$

Lesson 3–1

fact family

28
4   7

$4 \times 7 = 28, 7 \times 4 = 28,$
$28 \div 7 = 4, 28 \div 4 = 7$

Lesson 3–1

factor

$2 \times 2 = 4$

Lesson 3–5

Identity Property of Multiplication

$1 \times 10 = 10 \qquad 10 \times 1 = 10$

Ideas for Use

- Group 2 or 3 common words. Add a word that is unrelated to the group. Then work with a friend to name the unrelated word.

- Design a crossword puzzle. Use the definitions for the words as the clues.

- -

The order in which two numbers are multiplied does not change the product.

How is this property different from the Associative Property of Multiplication?

_____

_____

_____

The grouping of the factors does not change the product.

How is this property like the Associative Property of Addition?

_____

_____

_____

A number that is being divided.

Write the words from this set of cards that are related to *dividend*.

_____

To break apart a number.

Decompose is a multiple-meaning word. Use another meaning of *decompose* in a sentence.

_____

_____

A group of related facts using the same numbers.

How can you remember that a fact family uses the same 3 numbers?

_____

_____

_____

The number by which the dividend is being divided.

The suffix *-or* means "a thing that does the action". Write how *divisor* does the action in division.

_____

_____

When any number is multiplied by 1, the product is that number.

Look in the dictionary for a meaning of *identity*. Write a sentence using that meaning.

_____

_____

_____

A number that divides a whole number evenly. Also, a number that is multiplied by another number.

How can factors help you solve multiplication and division problems?

_____

_____

MY Vocabulary Cards

Processes & Practices

Lesson 3-7

multiple

multiples of 8:

0,   8,   16,   24,   32...

0 × 8   1 × 8   2 × 8   3 × 8   4 × 8

Lesson 3-1

product

3 × 4 = 12

Lesson 3-1

quotient

15 ÷ 3 = 5

Lesson 3-2

repeated subtraction

Lesson 3-5

Zero Property of Multiplication

12 × 0 = 0   3 × 0 = 0

Ideas for Use

- Write a tally mark on each card every time you read or write the word. Challenge yourself to use at least 3 tally marks for each card.

- Use the blank cards to write your own vocabulary cards.

The answer to a multiplication problem.

Product is a multiple-meaning word. Use another meaning of *product* in a sentence.

_____

_____

A multiple of a number is the product of that number and a whole number.

Write a clue to help you remember that multiples are used in multiplication.

_____

_____

Subtraction of the same subtrahend over and over again.

If repeated addition is a way of doing multiplication, what is repeated subtraction?

_____

The answer to a division problem.

Write a division equation whose quotient is 4.

_____

When any number is multiplied by 0, the product is 0.

Write a tip to help you remember this property.

_____

_____

_____

MY Foldable

Multiples

4

5

6

12

15

24

27

Factors

4

5

6

12

15

24

27

1

2

3

Multiples

Factors

Name ....

ESSENTIAL QUESTION
How are multiplication and division related?

You can use models to represent multiplication and division. Multiplication and division are opposite, or inverse, operations.

 Math in My World Tools Watch Tutor

High score!

Example 1

Felix and Mei-Ling went to an arcade. They played 4 games that cost $3 each. What was the total cost?

Write related multiplication and division sentences to solve.

1 Arrange counters in an array with 3 rows and 4 columns. Draw the counters in the chart.

There are _____ counters total.

2 Write a multiplication sentence.

rows   columns   total

□ × □ = □

factor   **factor**   **product**

3 Write a related division sentence.

total   rows   columns

□ ÷ □ = □

dividend   **divisor**   **quotient**

So, the total cost was $ _____ .

A **fact family** is a set of four related multiplication and division facts that use the same three numbers.

Example 2

Latanya and her father made an array of buttons. Write a fact family for the array.

There are 3 rows, 4 columns, and a total of 12 buttons.

$$\boxed{} \times \boxed{} = \boxed{} \qquad \boxed{} \div \boxed{} = \boxed{}$$

$$\boxed{} \times \boxed{} = \boxed{} \qquad \boxed{} \div \boxed{} = \boxed{}$$

Example 3

Vanessa has 36 books to put on 4 shelves. The same number of books will be placed on each shelf. How many books will be on each shelf?

Find _____ ÷ _____. You can use a related multiplication fact to help you divide.

$36 \div 4 = \blacksquare$ ⟵ (Find the unknown.)

THINK: $4 \times \blacksquare = 36$

$$4 \times \boxed{} = 36$$

So, $36 \div 4 = \boxed{}$.

Vanessa will place $\boxed{}$ books on each shelf.

Guided Practice

Write the fact family for each array.

1.

_____

_____

_____

2.

_____

_____

_____

Talk MATH

How are multiplication and division related?

Independent Practice

Write the fact family for each array or set of numbers.

3.

_____

_____

_____

4.

_____

_____

_____

5. 6, 9, 54

_____

_____

_____

_____

6. 7, 8, 56

_____

_____

_____

_____

7. 9, 11, 99

_____

_____

_____

_____

8. 11, 12, 132

_____

_____

_____

_____

Find each unknown to complete each fact family.

9. $4 \times 8 =$ _____

_____ $\times 4 = 32$

$32 \div$ _____ $= 8$

$32 \div 8 =$ _____

10. _____ $\times 9 = 72$

$9 \times 8 =$ _____

$72 \div$ _____ $= 8$

$72 \div 8 =$ _____

11. **Processes &Practices** **Model Math** Five banjo players enter the banjo contest. They are each playing four-string banjos. How many strings do the banjos have in all?

$5 \times 4 =$ _____

12. Ed wants to share 18 grapes equally among himself and two friends. How many grapes will each person receive?

$18 \div 3 =$ _____

Brain Builders

13. A Seminole basket-maker made a small basket. He used 9 blades of sweet grass and braided them for each coil. He used 81 blades in all. How many coils of sweet grass did he use to make the basket? Write a number sentence with an unknown, then solve.

_____

14. **Processes &Practices** **Use Number Sense** Draw an array using 24 dots. Write a fact family for your array.

_____

_____

15. **Building on the Essential Question** How can fact families and multiplication facts help you divide? Explain.

_____

_____

_____

Name ....

MY Homework

Homework Helper

Need help? connectED.mcgraw-hill.com

Write a fact family for the array.

> There are 2 rows, with 4 in each row. That makes a total of 8. The numbers in the fact family are 2, 4, and 8.

$2 \times 4 = 8$ $8 \div 2 = 4$

$4 \times 2 = 8$ $8 \div 4 = 2$

Find 14 ÷ 2. Use a related multiplication fact.

Think: $2 \times \blacksquare = 14$

        $2 \times 7 = 14$

> Using the related multiplication fact $2 \times 7 = 14$, you know that $14 \div 2 = 7$.

So, $14 \div 2 = 7$.

Practice

Write a fact family for each set of numbers.

1. 3, 6, 18

2. 2, 5, 10

_____

_____

_____

_____

_____

_____

Problem Solving

3. Monique volunteered 24 hours last month at the animal shelter. If she volunteered the same amount of hours each week for 4 weeks, how many hours did she volunteer each week?

4. **Processes &Practices** **Model Math** Tyler took 36 pictures while he was on vacation. He wants to put them into a photo album. He will put 6 pictures on each page. How many pages will he use?

Brain Builders

5. Lani's mother gave her and her two sisters $21 to spend at the movies. If each girl gets the same amount, how much do they each have to spend? Write related multiplication and division facts to solve.

Vocabulary Check

6. Use the vocabulary terms listed below to label each part of the equations.

dividend   divisor   factor   product   quotient

$$24 \div 4 = 6 \qquad\qquad 4 \times 6 = 24$$

Match each term with its definition.

7. division

- a group of related facts using the same numbers

8. fact family

- an operation on two numbers to find the product

9. multiplication

- an operation on two numbers to find the quotient

10. **Test Practice** Which is a related multiplication fact for $18 \div \blacksquare = 6$?

Ⓐ $18 \div 2 = 9$ Ⓑ $6 \times 3 = 18$ Ⓒ $18 - 12 = 6$ Ⓓ $6 \times 4 = 24$

Lesson 2
Relate Division and Subtraction

ESSENTIAL QUESTION
How are multiplication and division related?

You know that repeated addition can be used to multiply. **Repeated subtraction** can be used to divide.

 Math in My World ▶ Watch   💬 Tutor

Example 1

Wyatt is giving 15 colored pencils to 3 friends. How many colored pencils will each friend get?

You can use repeated subtraction to find 15 ÷ 3.

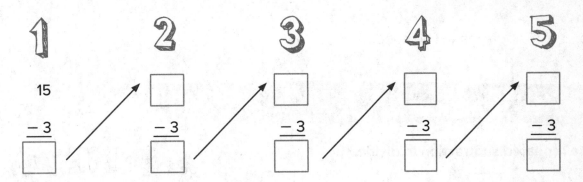

How many times was 3 subtracted from 15? _____

15 ÷ 3 = _____.

So, each friend will get _____ colored pencils.

Example 2

The students in Mr. Bantha's class are helping set up games for Family Fun Math Night. Each game can have four players. How many games will be needed for 12 people?

Find 12 ÷ 4.

You can skip count backward on a number line to find 12 ÷ 4.

 1 Find **12**.

2 Skip count backward by **4** until zero is reached.

3 Count the **number of times** 4 was subtracted.

The model shows that **12** − ☐ − ☐ − ☐ = ☐ .

Four was subtracted ☐ times.

12 ÷ 4 = 3

So, _____ games will be needed.

Guided Practice

Use repeated subtraction to divide.

1. 10 ÷ 2 = ☐

10 − 2 = ☐

☐ − 2 = ☐

☐ − 2 = ☐

☐ − 2 = ☐

☐ − 2 = ☐

2. 12 ÷ 3 = ☐

12 − 3 = ☐

☐ − 3 = ☐

☐ − 3 = ☐

☐ − 3 = ☐

Describe how to use subtraction to find 16 ÷ 4 without using a number line.

Independent Practice

Use repeated subtraction to divide.

3. $16 \div 8 = $ _____

4. $14 \div 2 = $ _____

5. $18 \div 6 = $ _____

6. $15 \div 5 = $ _____

7. $25 \div 5 = $ _____

8. $27 \div 9 = $ _____

9. $24 \div 8 = $ _____

10. $20 \div 4 = $ _____

11. $24 \div 6 = $ _____

Algebra Find each unknown number.

12. $12 \div 4 = \blacksquare$

13. $21 \div \blacksquare = 3$

14. $\blacksquare \div 5 = 2$

$\blacksquare = $ _____

$\blacksquare = $ _____

$\blacksquare = $ _____

Problem Solving

Write a number sentence to solve.

15. Mike buys a DVD set of a television series. Each DVD has 6 episodes. There are 24 episodes. How many DVDs are in the set?

_____

16. Lucy is making the same bracelet for 8 of her friends. She has 32 beads. How many beads can she put on each bracelet?

_____

Brain Builders

17. **Processes &Practices** **8** **Look for a Pattern** There are 9 mockingbird nests in a tree. There are a total of 18 eggs in the nests. Each nest has the same number of eggs. How many mockingbird eggs are in each nest?

_____

If there are only 6 nests, how many eggs are in each nest?

_____

18. **Processes &Practices** **3** **Find the Error** Olivia is using repeated subtraction to find $18 \div 2$. Find and correct her mistake.

$$18 - 2 - 2 - 2 - 2 - 2 - 2 - 2 - 2 = 2$$

19. **Building on the Essential Question** How are the operations of subtraction and division related? Explain.

_____

_____

_____

_____

Name

Homework Helper

Need help? connectED.mcgraw-hill.com

Find 12 ÷ 2. Use repeated subtraction.

$$
\begin{array}{cccccc}
12 & 10 & 8 & 6 & 4 & 2 \\
-2 & -2 & -2 & -2 & -2 & -2 \\
\hline
10 & 8 & 6 & 4 & 2 & 0 \\
\end{array}
$$

2 was subtracted 6 times

So, 12 ÷ 2 = 6.

Practice

Use repeated subtraction to divide.

1. 27 ÷ 3 = _____

2. 30 ÷ 10 = _____

3. 24 ÷ 6 = _____

4. 15 ÷ 1 = _____

5. 14 ÷ 7 = _____

6. 18 ÷ 3 = _____

7. 10 ÷ 5 = _____

8. 28 ÷ 4 = _____

9. 20 ÷ 4 = _____

Problem Solving

Write a number sentence for each situation. Then solve.

10. **Processes &Practices** **Model Math** A bag of 15 apples was shared among 5 boys equally. How many apples did each boy get?

_____

11. Cindy made 40 muffins. She placed them on 4 trays with the same amount on each tray. How many muffins did she put on each tray?

_____

Brain Builders

12. Shiloh put 24 eggs into 3 bowls with the same amount in each bowl. How many eggs did she put in each bowl?

_____

If there were 4 bowls, how many eggs would she put in each bowl?

_____

Vocabulary Check

13. Explain how you would use repeated subtraction to find $8 \div 2$.

_____

_____

_____

14. **Test Practice** Which number sentence is represented by the repeated subtraction at the right?

Ⓐ $9 \div 3 = 6$ Ⓑ $6 \div 3 = 2$

Ⓒ $9 \div 9 = 1$ Ⓓ $9 \div 3 = 3$

$$\begin{array}{ccc} 9 & 6 & 3 \\ -\,3 & -\,3 & -\,3 \\ \hline 6 & 3 & 0 \end{array}$$

Name ....

Multiplication as Comparison

Sometimes a problem uses a phrase like *times as many*, *times more*, and *times as much*. These kinds of problems are comparison problems.

Math in My World

Tools   Watch   Tutor

Example 1

Mary attended camp for 7 days this summer. Tyler attended 3 times as many days as Mary. Find the number of days Tyler attended camp.

Use counters to help you compare the groups of days.

1 Model Mary's days at camp as _____ group of 7 days. Draw your model.

2 Tyler attended 3 times as many days at camp.

Model Tyler's days at camp as _____ groups of 7. Draw your model.

3 Find the total of 3 groups of 7.

_____ + _____ + _____ = _____

or

_____ × _____ = _____

So, Tyler attended camp for _____ days.

A type of model drawing is the bar diagram. A bar diagram can help you understand a problem and plan to solve it.

Example 2

Suki used 15 beads to make a bracelet. This is 3 times as many beads as what Cassady used. How many beads did Cassady use?

 The bar diagram models this problem.

Suki used _____ times as many beads as Cassady.

 Find how many beads Cassady used. Write an equation.

$3 \times ? = 15$ $3 \times$ _____ $= 15$ beads   ← Use a fact family.

So, Cassady used _____ beads.

Guided Practice

1. Use multiplication or division to complete the equation for the phrase below.

   3 times as much

   $3 \times 3 = ?$

   $3 \times 3 =$ _____

One way to interpret $24 = 8 \times 3$ is to say that 24 is 8 times as many as 3. What is another way you can interpret this equation?

Name

_____

Independent Practice

Use multiplication or division to complete each equation and/or drawing.

2. 3 times as many

_____ × _____ = 3

3. 5 times more

_____ × _____ = 25

4. 4 times as much

_____ × 3 = 12

5. 10 times as much

[ ]

10 × _____ = 40

6. 2 times more

[ ]

2 × _____ = 6

7. twice as many

[ ]

2 × _____ = 14

Complete each bar diagram. Then complete the multiplication equation.

8. twice as many
as 4 boys

2 × _____ = 8

9. 2 times as many
as 3 bows

_____ × _____ = _____

10. 4 times as many
as 6 fish

_____ × _____ = _____

Draw a bar diagram. Then write a multiplication equation.

11. 3 times as much money as $6

12. 5 times as many as 1 star

Write the equation.

_____

Write the equation.

_____

Problem Solving

Draw a bar diagram and write an equation to solve.

13. **Processes &Practices** 4 **Model Math** There are 3 times as many blue balloons as green balloons. There are 4 green balloons. How many blue balloons are there?

14. Nan needs 4 times as much flour as sugar. She needs 4 cups of sugar. How much flour does she need?

Find the unknown.

_____ is 3 times as many as 4.

Find the unknown.

_____ is 4 times as many as 4.

 Brain Builders

15. **Processes &Practices** 2 **Use Number Sense** Circle the example that does not belong. Explain your reasoning.

$4 + 4 + 4 = 12$

$12 - 4 = 8$

16. **Building on the Essential Question** How can a bar diagram help me plan and solve a problem? Explain.

_____

_____

Name ....

Homework Helper

Need help? connectED.mcgraw-hill.com

Write a multiplication equation to describe the model.

5 times more

You need to find the total of 5 groups of 6.

Write a multiplication equation.   You can also show this in a bar diagram.

$6 \times 5 = 30$

Practice

Write a multiplication equation to describe each model.

1. 4 times more   **2.** 2 times as much   **3.** 6 times as many

_____   _____   _____

4. twice as many   **5.** 3 times more   **6.** 5 times as much

_____   _____   _____

7. 4 times as much money as $5

8. 2 times as many books as 7 books

4 × _____ = _____   _____ × _____ = _____

 # Problem Solving

9. Henry has 3 gerbils. Hannah has 3 times as many. How many gerbils does Hannah have?

Brain Builders

10. Dani needs 6 times as many red beads as gold beads. She needs 7 gold beads. How many red beads does Dani need? Explain how to model this problem.

11. Draw a bar diagram to represent 6 times more than $4. Now, show it another way, such as in an array.

12. Test Practice Amelia swam four times as many laps as Wesley. Wesley swam 7 laps. How many laps did Amelia swim?

ⓐ 11 laps   ⓒ 35 laps

ⓑ 28 laps   ⓓ 21 laps

Name ....

Lesson 4
Compare to Solve Problems

Additive comparisons use addition or subtraction to compare. Multiplicative comparisons use multiplication or division to compare.

Additive Comparison
how much more
how many more
how much less

Multiplicative Comparison
how many times more
how many times greater

 ## Math in My World

Surf's up!

Example 1

Bryan has been to a water park 4 times. Sarah has been to the water park three times as many times as Bryan. How many times has Sarah been to the water park?

Write an equation to find the unknown. You can use a letter, or a variable, to represent the unknown.

$4 \times$ _____ $= s$ ← the unknown: the number of times Sarah has been to the water park

Draw a picture to show 3 times as many as 4, or 3 groups of 4.

The picture shows a total of 12.

So, $4 \times 3 = 12$.

Since $s =$ _____ , Sarah has been to the

water park _____ times.

Example 2

Jess has 18 baseball cards. She has 6 times as many baseball cards as Max. How many baseball cards does Max have?

Write an equation. Let ■ represent the number of baseball cards that Max has.

■ × _____ = _____

Since ■ × 6 = 18, you know that 18 ÷ 6 = ■. ← [Use a fact family.]

Draw 6 equal groups of squares. Draw a total of 18 squares.

There are _____ squares in each group.

■ = 3. So, Max has _____ baseball cards.

Guided Practice

1. Sue has 3 times as many beads as Jo. Jo has 8 beads. How many beads does Sue have? Write an equation to find the unknown number. Use a variable for the unknown.

   3 × _____ = b

   b = _____

How can unknown numbers be represented in equations?

Independent Practice

Algebra **Write an equation to find the unknown number.**
Use a symbol for the unknown.

2. Paul drew 4 times as many pictures as Dennis. Paul drew 16 pictures. How many pictures did Dennis draw?

3. Maria made 21 cupcakes. This is three times as many as the number of cupcakes that Mary made. How many cupcakes did Mary make?

Algebra **Write an equation to find the unknown number.**
Use a variable for the unknown.

4. Wendy has dance class 2 days a week. James has dance class 5 times a week. How many more times a week does James have dance class than Wendy?

5. There are 4 fewer white bunnies than gray bunnies. There are 9 white bunnies. How many gray bunnies are there?

Use the table for Exercises 6–7.

6. How many more shoes were sold than belts?

7. Which item sold 2 times as many as shirts?

Items Sold at a Department Store	
Item	**Number Sold**
hats	4
shoes	7
belts	2
shirts	8
pants	16
socks	12

8. Which exercises on this page used addition or subtraction to compare? List them.

9. Which exercises on this page used multiplication or division to compare? List them.

Problem Solving

10. Jerry read 24 pages this weekend. That is 4 times as many pages as what he read last weekend. How many pages did he read last weekend?

11. A bean plant is 3 inches tall. A corn plant is five times as tall. How many inches tall is the corn plant?

12. **Processes &Practices** **1** **Plan Your Solution** Hannah used 10 more blocks than Steve. Steve used 7 blocks. How many blocks did Hannah use?

Brain Builders

13. There are 10 fewer robins than cardinals. There are 3 times as many bluebirds as robins. There are 16 cardinals. How many bluebirds are there?

14. **Processes &Practices** **1** **Check for Reasonableness** Missy earned three times as many points as Jerry. Kimmie earned 9 more points than Jerry. Missy earned 21 points. How many points did Kimmie earn? Explain how you know the answer is reasonable.

15. **Building on the Essential Question** How can you tell the difference between additive comparison and multiplicative comparison?

Name

MY Homework

Homework Helper

Need help? connectED.mcgraw-hill.com

Additive comparisons use addition or subtraction to compare.

Multiplicative comparisons use multiplication or division to compare.

Maya went swimming 7 times this month. Her brother went swimming 14 times this month. How many times more did Maya's brother go swimming than Maya?

Write an equation. Let b represent the unknown.

$7 \times b = 14$

$7 \times 2 = 14$

So, $b = 2$.

Maya's brother went swimming 2 times as much as Maya.

Additive Comparison
how much more
how many more
how much less

Multiplicative Comparison
how many times more
how many times greater

Practice

Algebra Write an equation to find the unknown number. Use a symbol for the unknown.

1. Julie earned $25. This is 5 times as much as what Lisa earned. How much did Lisa earn?

2. The red team scored 4 goals. The blue team scored 3 times as many goals. How many goals did the blue team score?

Problem Solving

**Write an equation to find the unknown number.
Use a variable for the unknown.**

3. A large aquarium has 6 more fish than a small aquarium. There are 19 fish in a large aquarium. How many fish are in a small aquarium?

Brain Builders

Processes &Practices **1** **Plan Your Solution** The table shows the number of roller coasters each student rode at the amusement park.

4.

Riding Roller Coasters	
Student	Number of Roller Coasters
Sarah	?
Tom	15
Val	3
Warren	9

How many more roller coasters has Tom been on than Val?

Sarah rode a roller coaster twice as many times as Warren. How many times did Sarah ride?

5. **Test Practice** Which shows 7 times as many as 5?

Ⓐ 2

Ⓑ 5

Ⓒ 12

Ⓓ 35

Check My Progress

Vocabulary Check

1. Write each word from the word bank on the correct label.

dividend   divisor   fact family

factor   product   quotient

```
              ┌──────────┐
              │          │
              └──────────┘
           ┌──────┴──────┐
   3 × 2 = 6              6 ÷ 3 = 2
   2 × 3 = 6              6 ÷ 2 = 3
   ↑  ↑  ↑              ↑   ↑   ↑
┌─────────┐      ┌─────────┐ ┌─────────┐
│         │      │         │ │         │
└─────────┘      └─────────┘ └─────────┘
    ┌─────────┐      ┌─────────┐
    │         │      │         │
    └─────────┘      └─────────┘
```

Concept Check

Use repeated subtraction to divide.

2. 18 ÷ 6 = _____

3. 28 ÷ 7 = _____

Write the fact family for each set of numbers.

4. 6, 4, 24

5. 7, 6, 42

6. 8, 4, 32

Problem Solving

7. Jerry made lemonade with the lemons shown.

Write a fact family for the array of lemons.

_____

8. There are 21 fish in an aquarium at the zoo. In order to clean the aquarium, the zookeeper must put the fish in three smaller tanks. If the same number of fish are in each tank, how many fish are in each tank? Use repeated subtraction to solve.

_____

Brain Builders

9. **Algebra** Gina has 5 pencils. Her sister has twice as many pencils. Write both a multiplication and related division equation to find the unknown number. Use a variable for the unknown. Solve.

_____

10. **Test Practice** Find the missing number. 15 ÷ 3 = ■.

Ⓐ 1   Ⓒ 5

Ⓑ 3   Ⓓ 12

Name ....

Lesson 5
Multiplication Properties and Division Rules

ESSENTIAL QUESTION
How are multiplication and division related?

Math in My World

Jenny	Cliff
Pack lunches	Set table
Take out trash	Clean room
	Walk dog

Example 1

The table shows Jenny's and Cliff's chores. Jenny earns $3 for each chore and Cliff earns $2 for each chore. How much does each person earn for completing chores?

Complete each number sentence.

Jenny

Cliff

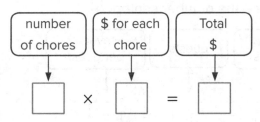

So, each person earns _____. The order in which the factors are multiplied does not change the product.

Key Concept Multiplication Properties

Commutative Property of Multiplication
When multiplying, the order of the factors does not change the product.

$4 \times 2 = 8$
$2 \times 4 = 8$

Identity Property of Multiplication
When any number is multiplied by 1, the product is that number.

$4 \times 1 = 4$

Zero Property of Multiplication
When any number is multiplied by 0, the product is 0.

$3 \times 0 = 0$

Helpful Hint

You will learn about the Associative Property of Multiplication and the Distributive Property in later lessons.

Copyright © McGraw-Hill Education

Online Content at connectED.mcgraw-hill.com

Lesson 5 161

The following rules can help you with division.

Key Concept Division Rules

Zeros in Division
When you divide 0 by any nonzero number, the quotient is 0.

$0 \div 5 = 0$

It is not possible to divide a number by 0.

Ones in Division
When you divide any number by 1, the quotient is always the dividend.

$7 \div 1 = 7$

When you divide any nonzero number by itself, the quotient is always 1.

$4 \div 4 = 1$

Helpful Hint

Quotient is the answer to a division problem.

Dividend is the number being divided.

Watch   Tutor

Example 2

There are 9 party favors and 9 guests. How many party favors will each guest get?

Complete the number sentence.

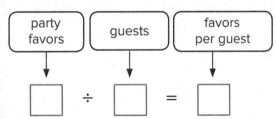

A nonzero number divided by the same number is 1.

So, each guest will get _____ party favor.

Talk MATH

Explain why the Identity Property of Multiplication uses 1 while the Identity Property of Addition uses 0.

Guided Practice

Identify the property or rule shown by each equation.

1. $12 \times 0 = 0$

2. $8 \times 5 = 5 \times 8$

_____ _____

_____ _____

Independent Practice

Identify the property or rule shown by each equation.

3. $6 \div 1 = 6$

4. $10 \div 10 = 1$

5. $8 \times 0 = 0$

6. $0 \div 12 = 0$

7. $22 \times 1 = 22$

8. $4 \times 3 = 3 \times 4$

Algebra Find each unknown number. Identify the property or rule.

9. $3 \div \blacksquare = 1$

$\blacksquare = \underline{\hspace{2cm}}$

10. $\blacksquare \times 8 = 8 \times 4$

$\blacksquare = \underline{\hspace{2cm}}$

11. $\blacksquare \div 11 = 0$

$\blacksquare = \underline{\hspace{2cm}}$

12. $\blacksquare \times 1 = 15$

$\blacksquare = \underline{\hspace{2cm}}$

Problem Solving

13. On a hiking trip, Tamika and Bryan hiked 7 miles a day. They hiked for 5 days. Kurt and Sade hiked 5 miles a day. How many days did it take Kurt and Sade to hike the same distance as Tamika and Bryan? Write a number sentence to solve.

14. There are 6 shelves in the library that each have 8 books displayed. How many books are on all of the shelves? Use the Commutative Property to write the multiplication sentence in two ways. Then solve.

Brain Builders

15. **Processes &Practices** **7** **Identify Structure** How would you describe the Identity Property of Multiplication? Use an example to show why it makes sense.

16. **Processes &Practices** **1** **Make a Plan** Write a multiplication problem that uses the Commutative Property of Multiplication to solve. Include a solution for your problem.

17. **Building on the Essential Question** How do multiplication properties and division rules help you to multiply and divide? Give an example.

Name

MY Homework

Homework Helper

Need help? ↗ connectED.mcgraw-hill.com

The tables show the properties of multiplication and the division rules that can be used to help solve problems. Identify the property or rule used in the equation $5 \times 1 = 5$.

Multiplication Properties	
Commutative Property of Multiplication When multiplying, the order of the factors does not change the product.	$3 \times 4 = 12$ $4 \times 3 = 12$
Identity Property of Multiplication When any number is multiplied by 1, the product is that number.	$7 \times 1 = 7$
Zero Property of Multiplication When any number is multiplied by 0, the product is 0.	$6 \times 0 = 0$

Division Rules	
Zeros in Division When you divide 0 by any nonzero number, the quotient is 0. It is not possible to divide a number by 0.	$0 \div 9 = 0$
Ones in Division When you divide any number by 1, the quotient is always the dividend. When you divide any nonzero number by itself, the quotient is always 1.	$8 \div 1 = 8$ $6 \div 6 = 1$

The equation $5 \times 1 = 5$ shows the Identity Property of Multiplication.

Practice

Identify the property or rule shown by each equation.

1. $9 \div 1 = 9$

2. $33 \times 1 = 33$

Problem Solving

Complete each number sentence. Identify the property or rule.

3. $5 \div$ _____ $= 5$

_____

4. $9 \times 8 = 8 \times$ _____

_____

5. _____ $\div 12 = 0$

_____

Brain Builders

6. **Processes &Practices** **7** **Identify Structure** Dennis has 3 packs of pens with 2 pens in each pack. He has 2 packs of pencils with 3 pencils in each pack. Write two multiplication sentences to show how many pens and pencils he has. Does he have more pens or more pencils? Explain.

_____

Vocabulary Check

Write a number sentence for each rule or property.

7. Ones in Division _____

8. Commutative Property of Multiplication _____

9. Zeros in Division _____

10. Zero Property of Multiplication _____

11. Identity Property of Multiplication _____

12. **Test Practice** The Zero Property of Multiplication tells you that 25×0 is equal to what number?

(A) 0   (B) 1   (C) 7   (D) 25

Name ....

ESSENTIAL QUESTION
How are multiplication and division related?

The **Associative Property of Multiplication** shows that the way in which numbers are grouped does not change their product.

 Math in My World Watch Tutor

Example 1

There are 2 video games in each value pack. There are 6 value packs in each box. If Raul buys 3 boxes for his collection, how many video games will he have?

You need to find $2 \times 6 \times 3$. There are two ways to group the numbers.

One Way	**Another Way**
Multiply 2 × 6 first.	**Multiply 6 × 3 first.**

$2 \times 6 \times 3 = (2 \times 6) \times 3$

$= \boxed{} \times \boxed{}$

$2 \times 6 \times 3 = 2 \times (6 \times 3)$

$= \boxed{} \times \boxed{}$

Helpful Hint
Use repeated addition to find 12×3.

$12 + 12 + 12 = \boxed{}$

$= \boxed{}$

Helpful Hint
Use repeated addition to find 18×2.

$18 + 18 = \boxed{}$

$= \boxed{}$

So, Raul will have _____ video games.

Example 2

Use the Associative Property of Multiplication to find 9 × 2 × 4.

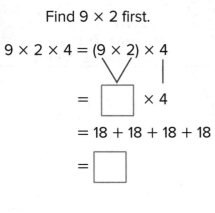

Find 9 × 2 first.

$9 \times 2 \times 4 = (9 \times 2) \times 4$

$= \boxed{} \times 4$

$= 18 + 18 + 18 + 18$

$= \boxed{}$

It is easier to find 9 × 8 than 18 × 4.

Find 2 × 4 first.

$9 \times 2 \times 4 = 9 \times (2 \times 4)$

$= 9 \times \boxed{}$

$= \boxed{}$

Helpful Hint

Parentheses ( ) tell you which numbers to multiply first.

Guided Practice

Multiply. Use the Associative Property.

1. $5 \times 3 \times 3 = \quad 5 \quad \times (3 \times 3)$

$= \boxed{} \times \boxed{}$

$= \boxed{}$

2. $4 \times 2 \times 7 = (4 \times 2) \times \quad 7$

$= \boxed{} \times \boxed{}$

$= \boxed{}$

3. $3 \times 1 \times 6 = (3 \times 1) \times \quad 6$

$= \boxed{} \quad \boxed{}$

$= \boxed{}$

Identify the order that makes it easier to multiply the factors in 9 × 4 × 2. Explain.

Name ....

Independent Practice

Multiply. Use the Associative Property.

4. $6 \times 1 \times 5 =$ _____

5. $2 \times 2 \times 7 =$ _____

6. $7 \times 5 \times 2 =$ _____

7. $10 \times 2 \times 5 =$ _____

8. $9 \times 3 \times 3 =$ _____

9. $6 \times 2 \times 2 =$ _____

10. $2 \times 3 \times 7 =$ _____

11. $9 \times 2 \times 4 =$ _____

12. $5 \times 1 \times 10 =$ _____

Compare. Use >, <, or =.

13. $4 \times 2 \times 9 \bigcirc 7 \times 4 \times 2$

14. $6 \times 2 \times 6 \bigcirc 5 \times 2 \times 8$

Find the value of each number sentence if
★ = 2, ☺ = 3, and ★ = 4.

15. $5 \times 1 \times$ ★ $=$ _____

16. $6 \times$ ★ $\times 3 =$ _____

17. ☺ $\times 3 \times$ ★ $=$ _____

Algebra Find the unknown number.

18. $4 \times$ ■ $\times 1 = 12$

19. $2 \times 5 \times$ ■ $= 60$

20. ■ $\times 3 \times 4 = 24$

■ = _____

■ = _____

■ = _____

Problem Solving

Write a number sentence to solve.

21. Gabriel is training for a race. On days that he jogs, he jogs 2 miles. His exercise schedule is shown. How many miles will he jog in 6 weeks?

_____

_____

Exercise Schedule	
Day	Activity
Monday	jog
Tuesday	jog
Wednesday	basketball
Thursday	jog
Friday	basketball
Saturday	jog
Sunday	rest

22. **Processes &Practices** **4** **Model Math** Bianca bikes 2 miles to her grandfather's house and 2 miles back to her house 5 times each month. How many miles does she bike each month?

_____

_____

Brain Builders

23. **Processes &Practices** **2** **Reason** Circle the equation that does not belong with the other three. Explain. Write an equation that does belong.

| $4 \times \blacksquare \times 7 = 56$ | $5 \times 2 \times \blacksquare = 40$ | $\blacksquare \times 3 \times 9 = 54$ | $4 \times \blacksquare \times 5 = 40$ |

_____

24. **Building on the Essential Question** How does the Associative Property of Multiplication help you to calculate products mentally? Give an example.

_____

_____

_____

Name

MY Homework

Homework Helper

Need help? ↗ connectED.mcgraw-hill.com

Find **3 × 4 × 2.**

One Way

Multiply 3 × 4 first.

$3 \times 4 \times 2 = (3 \times 4) \times 2$

$= \quad 12 \quad \times 2$

$= \quad\quad 24$

Another Way

Multiply 4 × 2 first.

$3 \times 4 \times 2 = 3 \times (4 \times 2)$

$= 3 \times \quad 8$

$= \quad 24$

So, $3 \times 4 \times 2 = 24$.

Practice

Multiply. Use the Associative Property.

1. $5 \times 2 \times 7 = $ _____

2. $8 \times 3 \times 2 = $ _____

3. $4 \times 2 \times 5 = $ _____

4. $5 \times 4 \times 3 = $ _____

5. $8 \times 2 \times 2 = $ _____

6. $3 \times 2 \times 5 = $ _____

Algebra Find the unknown in each equation.

7. $4 \times $ _____ $\times 8 = 64$

8. $3 \times 4 \times $ _____ $= 120$

9. $4 \times 2 \times $ _____ $= 40$

10. $6 \times 2 \times $ _____ $= 96$

Problem Solving

11. A bus seats 4 people in a row, and there are 12 rows. How many people would fit on 2 buses?

_____

 Brain Builders

12. Lucy has 2 pet snakes. Each snake eats 2 mice twice a week. How many mice will the snakes eat in 6 weeks?

_____

_____

13. Tyler delivers newspapers. He collected $10 from each customer. He has 5 customers on each street, and he delivers to 4 different streets. How much money did Tyler collect? Explain how to use a property to solve.

_____

_____

Vocabulary Check

Draw a line to match each property with the equation that represents it.

14. Associative Property of Multiplication

· $8 \times 9 = 9 \times 8$

15. Commutative Property of Multiplication

· $(4 \times 7) \times 2 = 4 \times (7 \times 2)$

16. Test Practice Emerson runs 2 miles to her friend's house 3 times each week. How many miles does she run in 4 weeks?

(A) 9 miles

(C) 24 miles

(B) 10 miles

(D) 50 miles

Name ....

Lesson 7
Factors and Multiples

ESSENTIAL QUESTION
How are multiplication and division related?

 Math in My World Tools Watch Tutor

Example 1

Mrs. Navarro is arranging desks in her classroom. There are 12 desks. How many ways can she arrange the desks so that the number of desks in each row is the same?

To find the different arrangements of desks, break down or **decompose** 12 into its factors. Think of factors that result in a product of 12. Use factors to write a number sentence for the arrays shown.

□ × □ = □

□ × □ = □

□ × □ = □

> **Helpful Hint**
> There are three more possible arrays:
>
> 12 × □
>
> 6 × □
>
> 4 × □

The factors of 12 are ___, ___, ___, ___, ___, and ___ .

So, the desks can be arranged in ___ ways.

Copyright © McGraw-Hill Education ©Chris Ryan/OJO Images/age fotostock

Online Content at **connectED.mcgraw-hill.com**

Lesson 7 173

A **multiple** of a number is the product of that number and any whole number. For example, 15 is a multiple of 5 because it is composed, or made, of 3 groups of 5. 15 is also a multiple of 3.

Example 2

All of the numbers listed in Row 7 or Column 7 are multiples of 7. Shade the multiples of 7 on the chart.

The first five multiples of 7 are 0,

_____, _____, _____, and _____.

28 is a multiple of 7 because it is composed of _____ groups of _____.

×	0	1	2	3	4	5	6	7	8	9	10
0	0	0	0	0	0	0	0	0	0	0	0
1	0	1	2	3	4	5	6	7	8	9	10
2	0	2	4	6	8	10	12	14	16	18	20
3	0	3	6	9	12	15	18	21	24	27	30
4	0	4	8	12	16	20	24	28	32	36	40
5	0	5	10	15	20	25	30	35	40	45	50
6	0	6	12	18	24	30	36	42	48	54	60
7	0	7	14	21	28	35	42	49	56	63	70
8	0	8	16	24	32	40	48	56	64	72	80
9	0	9	18	27	36	45	54	63	72	81	90
10	0	10	20	30	40	50	60	70	80	90	100

Guided Practice

Find the factors of each number.

1. 6

2. 36

Explain how factors and multiples are related.

Identify the first five multiples.

3. 4

4. 9

Name

Independent Practice

Find the factors of each number.

5. 4

6. 7

7. 14

8. 28

9. 30

10. 35

List the first five multiples.

11. 1

12. 3

13. 5

14. 7

15. 8

16. 6

Tell the total number modeled by each array. Then find the factors of that number.

17.

18.

Problem Solving

19. Complete the Venn Diagram.

factors
of 28   factors
of both   factors
20 and 28 of 20

20. Pedro walks his dog 3 times a day. How many times does Pedro walk his dog in one week? Find multiples of 3 to tell how many times Pedro walks his dog in 8, 9, and 10 days.

Brain Builders

21. Processes &Practices **5** **Use Math Tools** There are 24 cans of soup on a shelf. One way the cans can be displayed is in a 1 × 24 array. Think of factors of 24 to identify all the ways the cans can be displayed.

22. Processes &Practices **1** **Make a Plan** Identify the two numbers less than 20 with the most factors. Explain how you solved this problem.

23. **?** **Building on the Essential Question** How do you know when you have found all the factors of a number? Give an example.

Name _____

Homework Helper

Need help? connectED.mcgraw-hill.com

Find the factors of 16.

Think of factors that result in a product of 16.

$1 \times 16 = 16$
$2 \times 8 = 16$
$4 \times 4 = 16$

So, the factors of 16 are 1, 2, 4, 8, and 16.

Identify the first five multiples of 4.

First multiple: $0 \times 4 = 0$
Second multiple: $1 \times 4 = 4$
Third multiple: $2 \times 4 = 8$
Fourth multiple: $3 \times 4 = 12$
Fifth multiple: $4 \times 4 = 16$

So, the first five multiples of 4 are 0, 4, 8, 12, and 16.

Practice

Find the factors of each number.

1. 14

_____

2. 20

_____

Identify the first five multiples for each number.

3. 2

_____

4. 3

_____

5. 6

_____

6. 5

_____

7. 8

8. 7

Problem Solving

9. Each music class sings 8 songs a week. How many songs does each class sing in 5 weeks? 6 weeks? 7 weeks?

10. **Processes & Practices** **Use Math Tools** Tyra wants to arrange 32 tiles in equal rows and columns. How many ways could she organize the tiles? List the factors.

Brain Builders

11. Members of the marching band line up in 6 rows of 8. How many total members are there? Identify all the other ways they could line up in equal rows and columns.

Vocabulary Check

Complete each sentence with the correct vocabulary term.

decompose   multiple

12. The number 12 is a _____ of 2, 3, and 4.

13. One way to _____ 24 is to show it as 2 × 12.

14. **Test Practice** Which set of numbers correctly shows all of the factors of 28?

　Ⓐ 1, 2, 4, 7, 14, 28   Ⓒ 1, 2, 7, 14, 28

　Ⓑ 0, 1, 7, 14, 28   Ⓓ 1, 2, 4, 7, 8, 14, 28

Lesson 8
Problem-Solving Investigation
STRATEGY: Reasonable Answers

Learn the Strategy

Tutor

Starr won 4 tickets at Family Game Night. Suzi won 5 times as many tickets. Is it reasonable to say that they won 24 tickets all together?

1 Understand

What facts do you know?

Starr won _____ tickets.

Suzi won _____ times as many tickets as Starr.

What do you need to find?

the total number of _____ won all together

2 Plan

Find 5 × 4. Then add that to 4.

3 Solve

Find 5 × 4.

Model 5 groups of 4.

So, 5 × 4 = 20. Suzi won 20 tickets. Add. 20 + 4 = _____

Compare. The girls won _____ tickets. The answer is reasonable.

4 Check

Does my answer make sense? Explain.

_____

Dasan delivers 283 newspapers each week. Lisa delivers 302 newspapers each week. Is 400 a reasonable estimate for the number of newspapers they deliver each week all together?

 Understand

What facts do you know?

_____

_____

What do you need to find?

_____

_____

 Plan

_____

_____

_____

 Solve

 Check

Does my answer make sense? Explain.

_____

_____

Apply the Strategy

Determine a reasonable answer for each problem.

1. The table shows the number of pennies collected by four children. Is it reasonable to say that Myron and Teresa collected about 100 pennies in all? Explain.

Pennies Collected	
Child	**Number of Pennies**
Myron	48
Teresa	52
Veronica	47
Warren	53

Brain Builders

2. **Processes & Practices** **Make Sense of Problems** Jay will make $40 per week doing yard work for 6 weeks. He is saving his money to buy camping equipment that costs $400. He has already saved $120. Is it reasonable to say that Jay will save enough money in 6 weeks? Explain.

_____

_____

_____

_____

3. Write a problem in which $180 would be a reasonable answer.

_____

_____

_____

Review Strategies

Use any strategy to solve each problem.

- Use the four-step plan.
- Check for reasonableness.

4. A truck holds the number of cars shown. A parking lot has 6 times as many cars. How many cars are in the parking lot?

_____

5. Processes &Practices ① **Check for Reasonableness** Jack's basketball games are 4 quarters that are each 8 minutes long. Is it possible for Jack to play 35 minutes in a game? Explain.

_____

6. Processes &Practices ③ **Find the Error** Mark and his dad are going to the amusement park. One roller coaster has 1,204 feet of track. Another roller coaster has 2,941 feet of track. Mark estimates that the two roller coasters have a total of about 3,000 feet of track. Find and correct his mistake.

_____

_____

_____

_____

7. It costs $12 for 2 admissions to miniature golf. Abby wants to invite 9 of her friends. At this rate, how much would it cost for 10 people?

_____

Name _____

Homework Helper

Need help? connectED.mcgraw-hill.com

Callie and Cole each used 24 craft sticks for their art project.
Dinah and Dennis each used 33 craft sticks for their project.
Is it reasonable to say they used about 100 craft sticks in all?

1 Understand

What facts do you know?

Callie and Cole each used 24 craft sticks.

Dinah and Dennis each used 33 craft sticks.

What do you need to find?

I need to determine if they used about 100 craft sticks in all.

2 Plan

I will round the number of craft sticks each pair of students
used. Then I will add the rounded numbers.

3 Solve

Round 24 to the nearest ten: 20   Round 33 to the nearest ten: 30

Add: $20 + 20 + 30 + 30 = 100$

So, it is reasonable to say the children used about 100 craft sticks in all.

4 Check

I will add the numbers.

$$\begin{array}{r} 24 \\ + 24 \\ \hline 48 \end{array} \qquad \begin{array}{r} 33 \\ + 33 \\ \hline 66 \end{array} \qquad \begin{array}{r} 48 \\ + 66 \\ \hline 114 \end{array}$$

114 is close to 100, so the estimate is reasonable.

Problem Solving

Determine a reasonable answer for each problem.

1. Kevin can carry a basket 5 feet. Rachel can carry it 3 feet farther than Kevin. Daniel can carry the basket half as far as Rachel. Is it reasonable to say that they can carry it 15 feet with each person taking only one turn?

2. Josh and Anthony have a lemonade stand. They charge $1 for 2 cups of lemonade. They sell 14 cups each afternoon. Is it reasonable to say Josh and Anthony earned more than $50 after 3 afternoons?

Brain Builders

3. Javier's grandmother lives 120 miles away. It takes 1 hour to travel 40 miles by train. It takes 1 hour to travel 60 miles by car. If Javier leaves at 7 A.M., is it reasonable to say he will arrive at his grandmother's by 9 A.M. if he takes the train? What if he drives by car?

4. **Processes &Practices** **1** **Make Sense of Problems** Brittany wants to make cookies for the whole fourth grade. One batch of dough makes 2 dozen cookies. There are 68 fourth graders at her school. Is it reasonable to say that Brittany will need more than two batches of dough? Explain.

Vocabulary Check

Write the letter of each definition on the line next to the correct vocabulary word.

1. **Associative Property of Multiplication** _____

2. **Commutative Property of Multiplication** _____

3. **decompose** _____

4. **dividend** _____

5. **divisor** _____

6. **fact family** _____

7. **factor** _____

8. **Identity Property of Multiplication** _____

9. **multiple** _____

10. **product** _____

11. **quotient** _____

12. **repeated subtraction** _____

13. **Zero Property of Multiplication**

A. A group of related facts using the same numbers.

B. The number by which the dividend is being divided.

C. The property that states that the order in which two numbers are multiplied does not change the product.

D. The property that states any number multiplied by zero is zero.

E. A number that is multiplied by another number.

F. The answer of a division problem.

G. The answer of a multiplication problem.

H. The product of a given number and any whole number.

I. The property that states that the grouping of the factors does not change the product.

J. A strategy that can be used to divide.

K. A number that is being divided.

L. The property that states when any number is multiplied by 1, the product is that number.

M. A way to break down a number into its factors.

Concept Check

Write a fact family for each set of numbers.

14. 3, 7, 21 _____ _____ _____ _____

15. 9, 5, 45 _____ _____ _____ _____

Use repeated subtraction to divide.

16. 42 ÷ 7 = _____   **17.** 56 ÷ 8 = _____   **18.** 36 ÷ 9 = _____

---

19. Use multiplication to complete the number sentence.

5 times as many

     $\boxed{} \times \boxed{} = \boxed{}$

Identify the property or rule shown by each equation.

20. $6 \times 8 = 8 \times 6$ **21.** $(3 \times 2) \times 6 = 3 \times (2 \times 6)$

_____   _____

_____   _____

Find the factors of each number.

22. 16   **23.** 18   **24.** 15

_____   _____   _____

List the first five multiples.

25. 2   **26.** 10   **27.** 12

_____   _____   _____

Name ....

Problem Solving

28. There are 8 cans of soup. There are 3 times as many cans of vegetables. How many cans of vegetables are there? Write an equation to find the unknown. Use a variable for the unknown.

_____

29. Claire has 15 green beads, 8 blue beads, and 4 yellow beads. If she puts them onto 3 strings equally, how many beads are on each string?

_____

Brain Builders

30. Two friends want to share the shells that they collected on their trip to the beach. They have 18 shells in all. Use related facts and draw an array that will help them decide how they can evenly divide their shells between two friends. If there were three friends, how many shells would each get?

Two friends get _____ shells.

Three friends get _____ shells.

31. If 7 people each ride a roller coaster 5 times, and it is $3 per person per ride, what is the total price they paid for the rides?

_____

32. **Test Practice** Marina has scored 9 points on each of 11 quizzes. How many points has she scored in all?

Ⓐ 9 points   Ⓒ 90 points

Ⓑ 20 points   Ⓓ 99 points

Reflect

Use what you learned about multiplication and division
to complete the graphic organizer.

Same

ESSENTIAL QUESTION

**How are multiplication
and division related?**

Different

Now reflect on the ESSENTIAL QUESTION Write your answer below.

_____

_____

_____

_____

Performance Task

Bright Ideas

Students in the Bright Ideas club sell various items. The table shows the items they sell and the number they ordered during October.

Show all your work to receive full credit.

Item	Number Ordered	Cost per Item
light bulbs	56	$8
box of nightlights	36	$3
box of solar lights	24	$6
box of LED lights	32	$15

Part A

The club orders an *additional* 3 cartons of solar lights. Each carton contains 6 boxes of lights. To find the additional cost, Mario first calculated 3 × 6. Then he multiplied the result by the cost of each box. Tia also calculated the total. She first found the cost of 6 boxes of lights. Then she multiplied the result by the number of cartons. Write a multiplication number sentence that shows that Mario and Tia's total are equal. Explain the property that the number sentence demonstrates.

Part B

Write a word statement that compares the cost of a box of LED lights to the cost of a box of nightlights. Then write a number sentence involving multiplication that shows the same comparison.

Part C

Light bulbs come in packages of 8. In order to find p, the number of packages in the order, Ashley writes the equation $56 \div 8 = p$. Write two *different* equations, one using multiplication and the other using division, that could also be used to find p. Then find the value of p.

Part D

The club ordered four times as many boxes of solar lights in October as they did in September. In the diagram, each light represents a box of solar lights. Show how you can find the number of boxes of solar lights the club ordered in September. Then state the number of boxes ordered in September.

Part E

The club members took all the boxes of nightlights to a fair to sell. They arranged the boxes on a table. The arrangement is shown viewed from above. Draw two other ways that the nightlights can be arranged so that each row has the same number of boxes.

☐☐☐☐☐☐☐☐☐
☐☐☐☐☐☐☐☐
☐☐☐☐☐☐☐☐☐
☐☐☐☐☐☐☐☐☐

Multiply with One-Digit Numbers

ESSENTIAL QUESTION

How can I communicate multiplication?

Let's Go Shopping!

Watch a video!

Brain Builders

MY Chapter Project

Healthful Foods Party

1. Your teacher will place you in a group and assign to your group a healthful food. Write that food in the space below.

Our group's healthful food: _____

2. Use the Internet or other nutrition resources to find the serving size of that food for someone your age, and the calories in one serving. Write them below.

Serving size: _____

Calories per serving: _____

3. Estimate how much of your healthful food is needed to serve the whole class for a healthful food party, and how many calories are in the serving for the whole class. Show your work. Write the estimates below.

Total amount: _____

Total calories: _____

Name ....

Multiply. Use models if needed.

1. $2 \times 3 =$ _____

2. $7 \times \$8 =$ _____

3. $9 \times 4 =$ _____

4. $\$7 \times 5 =$ _____

5. Evan's photo album has 8 pages of pictures. How many photos are in Evan's album if the same number of photos is on each page?

_____

Identify the place value of the highlighted digit.

6. 1,630

7. $5,367

8. 20,495

_____   _____   _____

Round each number to its greatest place value.

9. 26 _____

10. $251 _____

11. 4,499 _____

12. There are 1,366 students at Sunrise Elementary School. About how many students attend the school?

_____

Shade the boxes to show the problems you answered correctly.

How Did I Do? | 1 | 2 | 3 | 4 | 5 | 6 | 7 | 8 | 9 | 10 | 11 | 12 |

Online Content at connectED.mcgraw-hill.com

MY Math Words

Vocab
abc

Review Vocabulary

equation   factor   product

Making Connections

Use the review words to complete each section of the bubble map.
Write an example for or a sentence about each word.

MY Vocabulary Cards

Vocab
a_bc

Copyright © McGraw-Hill Education

Lesson 4–7

Distributive Property

$$8 \times 11 = (8 \times 10) + (8 \times 1)$$

Lesson 4–4

partial products

$$348 \times 6 = (300 + 40 + 8) \times 6$$
$$300 \times 6 = 1{,}800$$
$$40 \times 6 = 240$$
$$8 \times 6 = 48$$
$$1{,}800 + 240 + 48 = 2{,}088$$

Lesson 4–6

regroup

 → 130

Ideas for Use

The products of each place value are found separately, and then added together.

How is using partial products helpful?

_____

_____

_____

Multiply the addends of a number and then add the products.

What are the two operations you use when using the Distributive Property?

_____

To use place value to exchange equal amounts when renaming a number.

The prefix *re-* means "again." Describe another word with the same prefix.

_____

MY Foldable

FOLDABLES® Follow the steps on the back to make your Foldable.

Multiply without Regrouping

Multi-Digit Multiplication with Zeros

Multiply with Regrouping

Multiples

Copyright © McGraw-Hill Education

Area Model

30   2

3

$$\begin{array}{r} 32 \\ \times\ \underline{\ \ 3} \\ \hline \end{array}$$

$3,502 \times 6 = \underline{\hspace{2cm}}$

$7 \times 8 = \underline{\hspace{2cm}}$

$7 \times 80 = \underline{\hspace{2cm}}$

$7 \times 800 = \underline{\hspace{2cm}}$

$7 \times 8,000 = \underline{\hspace{2cm}}$

$7 \times \underline{\hspace{1cm}} = \underline{\hspace{2cm}}$

$$\begin{array}{r} 4,238 \\ \times\ \underline{\ \ 4} \\ \hline \end{array}$$

Name ....

Lesson 1
Multiples of 10, 100, and 1,000

ESSENTIAL QUESTION
How can I communicate
multiplication?

A multiple of a number is the product of that
number and any whole number. Any number that
is divisible by ten is a multiple of ten. You can use
multiples and number patterns to multiply.

 Math in My World

Example 1

**The whale shark is the world's largest fish. Its mouth is 5 feet long,
and each foot contains 600 teeth. How many teeth does a whale
shark have?**

Find 5×600. Use basic facts and patterns.
600 is a multiple of 10.

$5 \times 6 =$ _____ 5×6 ones $= 30$ ones $= 30$

$5 \times 60 =$ _____ 5×6 tens $= 30$ tens $= 300$

$5 \times 600 =$ _____ 5×6 hundreds $= 30$ hundreds $= 3,000$

So, a whale shark has _____ teeth.
Notice that the product is 5×6 with two zeros at the end.

Example 2

Find $3 \times 7,000$.

$3 \times 7 =$ _____ 3×7 ones $= 21$ ones $= 21$

$3 \times 70 =$ _____ 3×7 tens $= 21$ tens $= 210$

$3 \times 700 =$ _____ 3×7 hundreds $= 21$ hundreds $= 2,100$

$3 \times 7,000 =$ _____ 3×7 thousands $= 21$ thousands $=$ _____

So, $3 \times 7,000$ is _____.

Notice that the product is 3×7 with three zeros at the end.

When you know basic facts and number patterns,
you can multiply mentally.

Example 3

The weight of a fire truck is 8 × 4,000 pounds.
What is its weight in pounds?
Let w represent the weight.

Write an equation.

$w = 8 \times$ _____

You need to find 8 × 4,000.

$8 \times 4 =$ _____   8 × 4 ones = _____ ones = _____

$8 \times 40 =$ _____   8 × 4 tens = _____ tens = _____

$8 \times 400 =$ _____   8 × 4 hundreds = _____ hundreds = _____

$8 \times 4,000 =$ _____   8 × 4 thousands = _____ thousands = _____

Notice that the product is 8 × 4 with _____ zeros at the end.

Since 8 × 4,000 = _____ , $w =$ _____ .

So, the weight of the fire truck is _____ pounds.

Guided Practice

Multiply. Use basic facts and patterns.

1. 6 × 8 = 48

    6 × 80 = _____

    6 × 800 = _____

    6 × 8,000 = 48,000

2. 7 × 9 = 63

    7 × 90 = _____

    7 × 900 = _____

    7 × 9,000 = _____

Multiply. Use mental math.

3. 8 × 600 = _____

4. 9 × 9,000 = _____

Talk MATH

What is the product of 4 and 5,000? Explain why there are more zeros in the product than in the factors in the problem.

Independent Practice

Multiply. Use basic facts and patterns.

5. $5 \times 3 = 15$

$5 \times 30 = 150$

$5 \times 300 = \underline{\hspace{2cm}}$

$5 \times 3,000 = \underline{\hspace{2cm}}$

6. $3 \times 4 = 12$

$3 \times 40 = \underline{\hspace{2cm}}$

$3 \times 400 = \underline{\hspace{2cm}}$

$3 \times 4,000 = \underline{\hspace{2cm}}$

7. $8 \times 5 = \underline{\hspace{2cm}}$

$8 \times 50 = \underline{\hspace{2cm}}$

$8 \times 500 = \underline{\hspace{2cm}}$

$8 \times 5,000 = \underline{\hspace{2cm}}$

8. $9 \times 1 = \underline{\hspace{2cm}}$

$9 \times 10 = \underline{\hspace{2cm}}$

$9 \times 100 = \underline{\hspace{2cm}}$

$9 \times 1,000 = \underline{\hspace{2cm}}$

9. $3 \times 7 = \underline{\hspace{2cm}}$

$3 \times 70 = \underline{\hspace{2cm}}$

$3 \times 700 = \underline{\hspace{2cm}}$

$3 \times 7,000 = \underline{\hspace{2cm}}$

10. $6 \times 5 = \underline{\hspace{2cm}}$

$6 \times 50 = \underline{\hspace{2cm}}$

$6 \times 500 = \underline{\hspace{2cm}}$

$6 \times 5,000 = \underline{\hspace{2cm}}$

Multiply. Use mental math.

11. $4 \times 30 = \underline{\hspace{2cm}}$

12. $6 \times 40 = \underline{\hspace{2cm}}$

13. $7 \times 200 = \underline{\hspace{2cm}}$

14. $4 \times 500 = \underline{\hspace{2cm}}$

15. $3 \times 9,000 = \underline{\hspace{2cm}}$

16. $9 \times 6,000 = \underline{\hspace{2cm}}$

Algebra Use mental math to find the unknown numbers.

17. If $6 \times 7 = 42$,

then $6 \times \underline{\hspace{2cm}} = 4,200$.

18. If $5 \times 7 = 35$,

then $5 \times \underline{\hspace{2cm}} = 3,500$.

19. If $8 \times 3 = 24$,

then $8 \times \underline{\hspace{2cm}} = 2,400$.

20. If $2 \times 9 = 18$,

then $2 \times \underline{\hspace{2cm}} = 1,800$.

21. How many times greater is the product 4×300 than the product

4×30? $\underline{\hspace{2cm}}$

Problem Solving

22. Each ticket for a theme park is $30. What is the total cost for a family with 5 people?

_____

23. The cost for each person to eat for one week is $100. Find the total cost for a family of five to eat for one week.

_____

24. Suppose 5 friends each go on 70 rides. How many rides will they go on all together?

_____

Brain Builders

25. **Processes &Practices** **5** **Use Mental Math** Use mental math to order these products from least to greatest, 5 × 5,000, 5 × 50 and 5 × 500. Explain.

_____

_____

_____

26. **Processes &Practices** **4** **Model Math** Write two multiplication expressions that have a product of 20,000. Explain your thinking to a friend.

_____

_____

27. **Building on the Essential Question** Does the product of a multiple of 10 always have a zero in the ones place? Explain.

_____

_____

_____

Name ....

MY Homework

Homework Helper

Need help? connectED.mcgraw-hill.com

Find 7 × 5,000.

Use basic facts and patterns to find the product.

$7 × 5 = 35$ $7 × 5$ ones $= 35$ ones, or 35

$7 × 50 = 350$ $7 × 5$ tens $= 35$ tens, or 350

$7 × 500 = 3,500$ $7 × 5$ hundreds $= 35$ hundreds, or 3,500

$7 × 5,000 = 35,000$ $7 × 5$ thousands $= 35$ thousands, or 35,000

So, $7 × 5,000 = 35,000$.

Practice

Multiply. Use basic facts and patterns.

1. $4 × 1 =$ _____

   $4 × 10 =$ _____

   $4 × 100 =$ _____

   $4 × 1,000 =$ _____

2. $6 × 7 =$ _____

   $6 × 70 =$ _____

   $6 × 700 =$ _____

   $6 × 7,000 =$ _____

3. $3 × 6 =$ _____

   $3 × 60 =$ _____

   $3 × 600 =$ _____

   $3 × 6,000 =$ _____

4. $8 × 9 =$ _____

   $8 × 90 =$ _____

   $8 × 900 =$ _____

   $8 × 9,000 =$ _____

Multiply. Use mental math.

5. 2 × 70 = _____

6. 9 × 500 = _____

7. 7 × 4,000 = _____

8. 3 × 2,000 = _____

Algebra Find each unknown.

9. 30 × ■ = 120 _____

10. 6 × ■ = 3,600 _____

11. 2 × ■ = 800 _____

12. ■ × 600 = 7,200 _____

Problem Solving

13. Joe bought a house. His payments are $800 per month. How much will he pay in 5 months' time?

_____

Brain Builders

14. Erin wants to purchase 3 CDs for $10 each. She has $20. How much more money does she need? Explain how you know.

_____

15. Kamil makes $70 per week mowing lawns. How much will Kamil make in 6 weeks? Will he make more than $600 in 8 weeks? Explain.

_____

16. **Processes &Practices** **2** **Use Number Sense** Justin is planning to purchase books for $20 each month. How much money will he spend on books in two years?

_____

17. Test Practice Violet has 11 rolls of pennies. There are 50 pennies in each roll. How many pennies does Violet have all together?

   Ⓐ 5,500 pennies   Ⓒ 500 pennies

   Ⓑ 550 pennies   Ⓓ 55 pennies

Name ....

Lesson 2
Round to Estimate Products

To estimate products, you can round factors to their greatest place.

 Math in My World

Example 1

There are passenger trains in the world that can travel at 267 miles per hour. About how far could a train travel in 3 hours?

Estimate 3 × 267.

Round 267 to its greatest place value.

Then use basic facts and patterns to multiply.

3 × 267

3 × _____ = _____

Helpful Hint
267 rounds to _____.

To which place was 267 rounded? _____

Since 267 was rounded *up*, the estimated product is *greater* than the actual product.

So, the train can travel about _____ miles in _____ hours.

Example 2

Estimate 8 × 2,496.

Round 2,496 to the thousands place value. Then multiply using basic facts and patterns.

8 × 2,496

↓

8 × _____ = _____

Since 2,496 was rounded *down*, the estimated product is *less* than the actual product.

So, 8 × 2,496 is about _____.

Helpful Hint

2,496 rounds

to _____.

Example 3

Lacey's older brother is going to a four-year college. The cost of his tuition is $8,562 each year. About how much will 4 years of college tuition cost?

Estimate 4 × $8,562.
Round to the greatest place value. Then multiply.

4 × _____ = _____

Since 8,562 was rounded _____, the estimated

product is _____ than the actual product.

So, tuition will cost about _____.

College Tuition

Cost per Year

$8,562

Guided Practice

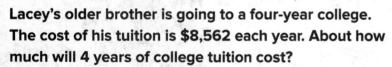

Estimate. Round to the greatest place value. Circle whether the estimate is *greater than* or *less than* the actual product.

1. 9 × $870

↓

9 × _____ = _____

greater than

less than

2. 3,293 × 3

↓

_____ × 3 = _____

greater than

less than

Which product is closer to the estimate of 1,600:

4 × 385 or 4 × 405? Explain.

Copyright © McGraw-Hill Education

Name _____

Estimate. Round to the greatest place value. Circle whether the estimate is *greater than* or *less than* the actual product.

3. 562 × 6   greater than   **4.** 2 × 896   greater than

                                        less than   less than

_____ × _____ = _____   _____ × _____ = _____

5. 729 × 8   greater than   **6.** 2 × $438   greater than

                                        less than   less than

_____ × _____ = _____   _____ × _____ = _____

7. $450 × 7   greater than   **8.** 3 × 5,489   greater than

                                        less than   less than

_____ × _____ = _____   _____ × _____ = _____

Draw lines to match each product with its most reasonable estimate.

9. 7 × 189   • 4,800

10. 211 × 9   • 1,400

11. 8 × 632   • 2,500

12. 455 × 5   • 1,800

Problem Solving

Toby and Lena like to go to the arcade. They earn points towards prizes. For Exercises 13–15, use the information to the right.

2,000   10,000   50,000   500

13. **Processes &Practices** ➊ **Plan Your Solution** Toby went to the arcade 2 times. He earned 5,150 points each time. What is the biggest prize Toby can get?

_____

14. How many toy cars could Toby get with his points?

_____

15. Lena went to the arcade 7 times. She earned 9,050 points each time. What are the two largest prizes she can get?

_____

_____

Brain Builders

16. The students in Mrs. Pluma's two classes wrote a total of 200 letters to their pen pals. Every student wrote 4 letters. How many students are in each class, if the classes are equal in size?

_____

17. **Processes &Practices** ➊ **Use Number Sense** Explain how you can tell if your estimated answer is greater or less than the exact answer to a multiplication problem.

_____

_____

18. **?** **Building on the Essential Question** How is estimation helpful when finding a product mentally? Explain.

_____

Name ....

MY Homework

Homework Helper

Need help? *connectED.mcgraw-hill.com*

Estimate 6 × 8,825.

1 Round the greater factor to its greatest place.

Helpful Hint

8,825 rounds up to 9,000

6 × 8,825

↓

→ 9,000

↓

6 × 9,000 = 54,000

2 Multiply. Use basic facts and patterns.

6 × 9,000 = 54,000

The estimate for 6 × 8,825 is 54,000.

Since 8,825 was rounded up, the estimate is greater than the actual product.

Practice

Estimate. Round to the greatest place value. Circle whether the estimate is *greater than* or *less than* the actual product.

1. 756 × 4

800 × 4 = _____

greater than   less than

2. $246 × 8

_____ × 8 = _____

greater than   less than

3. 4,528 × 4

_____ × 4 = _____

greater than   less than

4. 2,331 × 5

_____ × 5 = _____

greater than   less than

Estimate. Round to the greatest place value. Circle whether the estimate is *greater than* or *less than* the actual product.

5. 143 × 2

_____ × 2 = _____

greater than   less than

6. 2,721 × 4

_____ × 4 = _____

greater than   less than

7. 6 × $6,517

_____

greater than   less than

8. 7 × $9,499

_____

greater than   less than

Brain Builders

Estimate each product. Then write whether the estimate is *greater than* or *less than* the actual product.

9. There are 325 rows of 9 chairs in the theater. There are 102 chairs in the balcony. About how many chairs are there?

_____

10. Celine is estimating the number of tiles in a mosaic. There are 7 different colors, and 472 tiles of each color. There are also 96 blank tiles. What should Celine's estimate be?

_____

11. **Processes &Practices** 5 **Use Mental Math** A resort has 380 rooms with space for 6 people. Estimate the greatest number of people who could be in these rooms at the same time. Explain your thinking to a friend or family member.

_____

_____

12. **Test Practice** Kurt swims 575 yards at each swim practice. If he practices Monday, Wednesday, and Friday, about how many yards will he swim that week?

Ⓐ 150 yards   Ⓑ 180 yards   Ⓒ 1,500 yards   Ⓓ 1,800 yards

Name ....

Lesson 3
Hands On
Use Place Value to Multiply

ESSENTIAL QUESTION
How can I communicate multiplication?

You can use base-ten models to multiply by one-digit numbers.

Build It

Grace and her friends are at the mall. They see 3 rows of parked cars. There are 23 cars in each row. How many cars are there all together?

Find 3 × 23 using models.

1 Circle the ones. Count the number of ones.

_____ + _____ + _____ = _____

2 Circle the tens. Count the number of tens.

_____ + _____ + _____ = _____

3 × 23 = _____

So, Grace and her friends see _____ cars.

Write the number of ones in the ones place.

Ones		
hundreds	tens	ones

Write the number of tens in the tens place.

Copyright © McGraw-Hill Education   Digital Light Source, Inc.

Try It

Find 2 × 32 using models.

1 Model 2 × 32.

2 Circle the ones. Then count the ones. Complete the place-value chart.

There are _____ ones.

	Ones	
hundreds	tens	ones
	⬜	⬜

There are _____ tens.

3 Circle the tens. Count the number of tens. Complete the place-value chart.

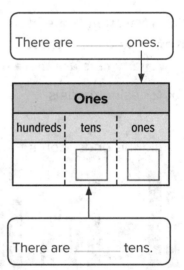

So, 2 × 32 = _____.

Talk About It

1. How would you model 2 × 22?

_____

2. **Processes &Practices** **3** **Justify Your Conclusion** Would you rather use base-ten blocks, counters, or mental math to model Activity 1? Explain.

_____

Practice It

Multiply. Use models. Draw your models.

3. $3 \times 22 =$ _____

4. $4 \times 12 =$ _____

5. $3 \times 20 =$ _____

6. $1 \times 56 =$ _____

Algebra **Find the unknown number. Use models. Draw your models.**

7. $4 \times 22 = a$

$a =$ _____

8. $2 \times 24 = c$

$c =$ _____

Apply It

Algebra Write an equation to solve.

9. **Use Symbols** There are 4 shoe stores at the local mall. There are 22 people in each shoe store. How many people are there in the four stores?

4 × _____ = _____

So, there are _____ people.

10. Justine earned $32 each month for 3 months. What is the total amount of money that she earned?

_____ × $32 = $ _____

So, she earned $ _____ .

11. Trent's cousin wants 2 tulips on each table at her wedding. There are 24 tables. How many tulips does she need?

_____ × _____ = _____

So, _____ tulips are needed.

12. **Use Number Sense** The product of two numbers is 88. The sum of the two numbers is 26. What are the numbers?

Write About It

13. How do models help you to multiply by one-digit numbers? Explain.

_____

_____

_____

Name

MY Homework

Homework Helper

Need help? connectED.mcgraw-hill.com

Use models to find 4 × 22.

 Count the number of ones.

2 + 2 + 2 + 2 = 8

Ones		
hundreds	tens	ones
		8

Count the number of tens.

20 + 20 + 20 + 20 = 80

Ones		
hundreds	tens	ones
	8	8

So, 4 × 22 = 88.

Practice

Multiply. Draw the models if needed.

1. 2 × 23 = _____

2. 4 × 21 = _____

3. 2 × 22 = _____

4. 3 × 11 = _____

Multiply. Draw the models if needed.

5. 3 × 32 = _____

6. 2 × 43 = _____

7. 4 × 12 = _____

8. 3 × 21 = _____

Problem Solving

Processes & Practices **Use Algebra Write an equation to solve.**

9. Julie hung 3 birdfeeders in her yard. Each birdfeeder has 12 perches. How many perches are there all together?

_____

10. There are 32 chairs in each classroom. What is the total number of chairs in 2 classrooms?

_____

11. Lance took 34 photos each day of his vacation. He was on vacation for 2 days. Write an equation that describes how many photos he took in all.

_____

12. There are 42 biscuits in each box. How many biscuits are in 2 boxes?

_____

13. Each prize is worth 3 tickets at the game center. How many tickets are needed for 23 prizes?

_____

Name _____

Lesson 4
Hands On
Use Models to Multiply

ESSENTIAL QUESTION
How can I communicate multiplication?

You can use **partial products** to multiply a one-digit number by a two-digit number. Find the products for the tens and ones separately. Then add them together.

Area models and arrays can show partial products.

Draw It

The hiking club has 3 groups of hikers. Each group has 12 people. How many hikers are in the hiking club all together?

Use an array to find 3 × 12.

1 Draw a rectangular array.
Separate 12 into

_____ and _____ .

2 Find the partial products.

3 × 10 = ☐

3 × 2 = ☐

3 Add the partial products.

☐ + ☐ = ☐

3 × 12 = _____

So, there are _____ hikers in the hiking club.

Try It

The hikers hiked for four hours. Each hour, they saw 21 animals. How many animals did they see in all?

Use an area model to find 4 × 21.

 Draw an area model. Separate 21 into

_____ and _____ . Label the model.

 Find the partial products.   4 × 20 = _____   4 × 1 = _____

 Add the partial products.   _____ + _____ = _____

So, 4 × 21 = _____ .

Talk About It

1. Explain how to draw an area model to represent 2 × 15.

_____

_____

_____

2. **Processes &Practices** **Explain to a Friend** Explain to a friend how you would multiply 3 × 32.

_____

_____

Practice It

Draw an array to multiply.

3. $3 \times 13 =$ _____

$3 \times 10 = 30$

$3 \times 3 = 9$

$30 + 9 =$ _____

4. $4 \times 12 =$ _____

5. $1 \times 26 =$ _____

Draw an area model to multiply.

6. $3 \times 31 =$ _____

7. $4 \times 22 =$ _____

_____ + _____ = _____

Algebra Find each unknown number. Use an array or area model.

8. $43 \times 2 = d$

$d =$ _____

9. $39 \times 1 = g$

$g =$ _____

Apply It

Use the table for Exercises 10–13. Draw models to solve.

Toy Sale		
Type of Toy	Regular Price	Sale Price
block sets	$20	$14
puzzles	$12	$10
action figure sets	$13	$12
cars	$11	$10

10. What is the total cost of 2 puzzles and 1 car at regular price?

_____

11. How much more are 2 block sets than 2 action figure sets at regular price?

_____

12. **Processes &Practices** 1 **Make Sense of Problems** How much is saved by buying 3 action figure sets at the sale price instead of the regular price?

_____

13. **Processes &Practices** 2 **Understand Symbols** Compare using the regular prices. Use >, <, or =.

4 cars + 2 action figure sets ◯ 4 puzzles + 1 block set

Write About It

14. How can I use area models to represent multiplication? Explain.

_____

_____

_____

Name ....

Homework Helper

Need help? connectED.mcgraw-hill.com

Hailey's parents had a garage sale. Hailey sold 24 games for $2 each. What is the total amount of money Hailey made?

Use an array to find $2 × 24. The array shows 2 rows of 24, which equals 48.

1 Find the partial products.

$2 × 20 = 40$

$2 × 4 = 8$

$\$2 × 24 = \$48$

So, Hailey made $48.

2 Add the partial products

$40 + 8 = 48$

Practice

Draw an array or area model to solve.

1. $3 × 22 =$ _____

2. $2 × 41 =$ _____

Draw an array or area model to solve.

3. $31 \times 3 =$ _____

4. $42 \times 2 =$ _____

5. $24 \times 2 =$ _____

6. $33 \times 2 =$ _____

 Problem Solving

Algebra Write an equation to solve.

7. Tyrone uses 2 cups of flour for each batch of cookies. How much flour will he need to make 31 batches of cookies?

_____

8. Corinne's puppy eats 3 meals a day. How many meals does the puppy eat in 32 days?

_____

9. Processes &Practices **Use Symbols** There are 3 shelves. Each shelf has 21 books. How many books are there in all?

_____

Vocabulary Check

10. Show how you would use partial products to solve 3×12.

Check My Progress

Vocabulary Check

> **1.** Match each definition to the corresponding vocabulary word.
>
> A(n) _____ of a number is the product of that number and any whole number.
>
> A number close to an exact value.
>
> A number that divides a whole number evenly. Also a number that is multiplied by another number.
>
> The products of each place value are found separately, and then added together.
>
> • estimate
>
> • factor
>
> • partial products
>
> • multiple

Concept Check

Multiply. Use basic facts and patterns.

2. $2 \times 60 = $ _____   **3.** $9 \times 600 = $ _____   **4.** $6 \times 4,000 = $ _____

Estimate. Round to the greatest place value. Circle whether the estimate is _greater than_ or _less than_ the actual product.

5. 6 × 423

6 × _____ = _____

greater than

less than

6. 1,987 × 5

_____ × 5 = _____

greater than

less than

Draw an array or area model to multiply.

7. 2 × 15 = _____

8. 3 × 19 = _____

Brain Builders

9. Grasshoppers can jump about 20 times their length. About how far could the grasshopper below travel in two jumps?

_____

10. There are 21 boxes of markers in each art room. Each box holds 6 markers. How many markers are in 2 art rooms?

_____

11. Test Practice To find the product of 2 × 200, Julia used the basic fact 2 × 2 = 4. How many zeros should she include in the product of 2 × 200?

Ⓐ 1 Ⓒ 3

Ⓑ 2 Ⓓ 4

Lesson 5
Multiply by a Two-Digit Number

Copyright © McGraw-Hill Education   Spotmatik/iStock/Getty Images

ESSENTIAL QUESTION
How can I communicate multiplication?

Place value can help you multiply.

 Math in My World   Tutor

Example 1

Ann's mom buys two helmets. Each helmet costs $24. How much does she spend on helmets?

You need to find 2 × $24.

Multiply.

1
$$\begin{array}{r} 24 \\ \times\ 2 \\ \hline 8 \end{array}$$
Multiply the ones.
2 × 4 ones = _____ ones

2
$$\begin{array}{r} 24 \\ \times\ 2 \\ \hline 48 \end{array}$$
Multiply the tens.
2 × 2 tens = _____ tens

Helpful Hint
Line up the factors by the ones place.

So, Ann's mom spends _____ .

Check for Reasonableness

The area model shows the partial products.

	20	4	
2	2 × 20 = 40	2 × 4 = 8	40 + 8 = ☐

So, the answer is correct.

You can estimate to check for reasonableness.

Example 2

Thirty-one bikes were ordered from Jimmy's Bike Shop. Each bike has two tires. How many tires will Jimmy need for the bikes?

Estimate $31 \times 2 \longrightarrow$ _____ $\times$ _____ = _____

1 **Multiply the ones.**

2×1 one $= 2$ ones

Write the product in the ones place.

2 **Multiply the tens.**

2×3 tens $= 6$ tens

Write the product in the tens place.

So, Jimmy will need _____ bike tires.

Check for Reasonableness

The product, _____, is close to the estimate, _____.

Guided Practice

Multiply. Check for reasonableness.

1.   42
   × 2
   ☐☐

2.   21
   × 3
   ☐☐

3. $11 \times 4 =$ _____

4. $32 \times 2 =$ _____

Suppose you found 99 as the product of 33×3. How can you check to see if your answer is reasonable?

Independent Practice

Multiply. Check for reasonableness.

5.
$$\begin{array}{r} 44 \\ \times\ 2 \\ \hline \end{array}$$
□ □

6.
$$\begin{array}{r} 21 \\ \times\ 4 \\ \hline \end{array}$$
□ □

7.
$$\begin{array}{r} 13 \\ \times\ 2 \\ \hline \end{array}$$
□ □

8. $41 \times 2 =$ _____

9. $12 \times 3 =$ _____

10. $4 \times 22 =$ _____

Algebra **Find each unknown number.**

11. $41 \times 2 = h$

$h =$ _____

12. $12 \times 3 = j$

$j =$ _____

13. $4 \times 22 = k$

$k =$ _____

Problem Solving

14. **Plan Your Solution** A city has 23 swing sets. Each has 3 swings. How many swings are there in all?

_____

15. The world's largest rodent is called a *capybara*. It can weigh 34 kilograms. How much could 2 capybaras weigh?

_____

Brain Builders

16. Recycling one ton of paper saves 17 trees. About how many trees are saved if 2 tons of paper are recycled the first month and 3 tons are recycled the second month?

_____

_____

17. **Processes &Practices** **6** **Explain to a Friend** Adrian has four boxes of action figures. There are 12 in each box. Alec has 21 action figures in each of his three boxes. Who has more action figures? Explain to a friend.

_____

_____

18. ❓ **Building on the Essential Question** How can estimation be used to check multiplication problems for reasonableness? Give an example.

_____

_____

_____

MY Homework

Homework Helper eHelp

Need help? connectED.mcgraw-hill.com

Find 4 × 11.

Estimate the product. 4 × 10 = 40

Multiply.

$$\begin{array}{r} 11 \\ \times\ 4 \\ \hline 44 \end{array}$$

1 **Multiply the ones.**
The product is in the ones place.

2 **Multiply the tens.**
The product is in the tens place.

So, 4 × 11 = 44.

Check for Reasonableness The product, 44, is close to the estimate, 40.

Practice

Multiply. Draw an area model. Check for reasonableness.

1. $\begin{array}{r} 30 \\ \times\ 2 \\ \hline \end{array}$

2. $\begin{array}{r} 21 \\ \times\ 4 \\ \hline \end{array}$

3. $\begin{array}{r} 86 \\ \times\ 1 \\ \hline \end{array}$

Multiply.

4. $3 \times 31 =$ _____   **5.** $6 \times 11 =$ _____

Problem Solving

6. Caroline makes $3 an hour pet-sitting for the neighbors. Last summer she worked 23 hours. How much money did Caroline earn?

_____

7. Simon has 12 CDs. He burns 3 copies of each. How many CDs did Simon make?

_____

Brain Builders

8. The school cafeteria has 4 rows of tables. Each row has 22 seats. The school patio has 12 tables. Each table seats 4 students. How many students can sit in the school cafeteria and on the patio at the same time?

_____

9. **Processes &Practices** **2** **Use Number Sense** Steve is playing a memory game with picture cards. He makes 2 boards. One board has 5 rows of 11 cards. The other board has 4 rows of 13 cards. Write a number sentence for the board with more cards.

_____

10. **Test Practice** John wants to buy birthday gifts for 4 friends. He can spend $20 for each gift. How much money can John spend on gifts in all?

   Ⓐ $80   Ⓒ $34

   Ⓑ $40   Ⓓ $24

Name ....

Lesson 6
Hands On
Model Regrouping

ESSENTIAL QUESTION
How can I communicate multiplication?

Build It

Sometimes you need to regroup to multiply numbers. Regrouping uses place value to exchange equal amounts when renaming a number.

Find 3 × 26.

1. Use base-ten blocks to model 3 × 26.

2. Find the ones.
   There are 18 ones. Regroup as 1 ten and 8 ones.
   Write an 8 in the ones place.

There are 8 ones.

3. Find the tens.
   Count the number of tens.
   Complete the place-value chart.

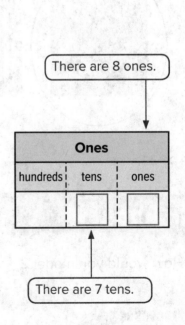

Ones		
hundreds	tens	ones

There are 7 tens.

So, 3 × 26 = _____.

Find 4 × 31.

1 Use base-ten blocks to model 4 × 31.
Find the ones. Count the ones. There are _____ ones.

2 Find the tens. There are 12 tens.
Regroup as 1 hundred and _____ tens.

3 Find the hundreds. Count the number of hundreds.
There is _____ hundred. Complete the place-value chart.

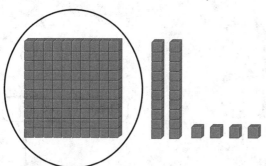

Ones		
hundreds	tens	ones

So, 4 × 31 = _____.

Talk About It

1. How would you model 2 × 38?

Processes
2. **&Practices** **Make a Plan** How would you model 4 × 52?

Name ....

Practice It

Multiply. Use models.

3. $2 \times 17 =$ _____

4. $4 \times 32 =$ _____

5. $3 \times 44 =$ _____

6. $4 \times 54 =$ _____

7. $3 \times 28 =$ _____

8. $4 \times 63 =$ _____

9. $2 \times 48 =$ _____

10. $6 \times 24 =$ _____

11. $4 \times 38 =$ _____

12. $5 \times 27 =$ _____

Apply It

13. Willow is shopping for a soccer ball. How many soccer balls *n*, are on the display?

$n =$ _____ × _____

There are _____ balls on the display.

14. Suppose there are 6 displays in the store. How many soccer balls *n*, are there in all?

$n =$ _____ × _____

There are _____ balls on 6 displays.

15. Processes &Practices 2 **Use Algebra** Each crate has 35 soccer balls. How many soccer balls *n*, are there in 4 crates?

$n =$ _____ × _____

There are _____ balls in 4 crates.

16. Processes &Practices 2 **Reason** Describe what would happen if you changed the number of balls on the display. How would that change the number of balls on 6 displays?

_____

_____

_____

Write About It

17. How is regrouping for addition similar to regrouping for multiplication?

_____

_____

_____

Name _____

MY Homework

Homework Helper   eHelp

Need help? connectED.mcgraw-hill.com

Find 4 × 14.

 Use base-ten blocks to model 4 × 14.

2 Find the number of ones. There are 16 ones. Regroup them as 1 ten and 6 ones.

Ones		
hundreds	tens	ones
		6

3 Find the number of tens. There are 5 tens.

Ones		
hundreds	tens	ones
	5	6

So, 4 × 14 = 56.

Practice

1. Multiply. Draw models if needed. 2 × 46 = _____

Multiply. Draw models if needed.

2. $3 \times 38 =$ _____

3. $4 \times 46 =$ _____

4. $5 \times 23 =$ _____

5. $6 \times 37 =$ _____

Problem Solving

6. **Processes &Practices** **2** **Use Algebra** Marissa is doing the laundry. She folds 16 items of clothing for each member of her family. There are 6 people in Marissa's family. How many items of clothing does she fold?

_____

7. Ben and his dad sold 9 dozen sunflowers at the farmer's market. There were 12 flowers in one dozen. What is the total number of sunflowers Ben and his dad sold?

_____

8. Riko rides her bike 26 miles every week. What is the total number of miles she will ride in 7 weeks?

_____

Vocabulary Check Vocab abc

9. Explain how to regroup 43 ones as tens and ones.

_____

Name _____

Lesson 7
The Distributive Property

The **Distributive Property** can be used to multiply greater numbers. It combines addition and multiplication. First, the numbers are decomposed, or broken down into parts. Then, the parts are multiplied separately and then added together.

 Math in My World [Watch] [Tutor]

Example 1

Chef Cora hard-boils six dozen eggs each day to make egg salad for sandwiches at the food court. How many eggs does she hard-boil each day?

There are 12 eggs in one dozen.

Find 6 × 12.

Decompose 12 into 10 + 2.

Think of 6 × 12 as (6 × 10) + (6 × 2).

6 × 12 = (___ × ___) + (___ × ___)   Decompose 12 into 10 + 2.

= ___ + ___   Find each product mentally.

= ___   Add the products.

So, Chef Cora hard-boils ___ eggs each day.

Key Concept   Distributive Property

Words	The Distributive Property says that you can multiply the addends of a number and then add the products.
Symbols	$6 \times 12 = (6 \times 10) + (6 \times 2)$

Example 2

Twenty-seven students are attending a performance at a children's theater. Each admission ticket is $5. What is the total cost for the 27 students?

Let c represent the total. Write an equation.

$c =$ _____ $\times 5$

Find 27×5.

$27 \times 5 = (20 \times 5) + (7 \times 5)$

$=$ _____ $+$ _____

$=$ _____

Since $27 \times 5 =$ _____, $c =$ _____.

So, the total cost for 27 students is _____.

> ### Talk MATH
> How can you use the Distributive Property or an area model to find 3×24?

Guided Practice

Use the Distributive Property to multiply. Draw an area model.

1. $12 \times 9 =$ _____

```
        10      +  2
     ┌───────┬────────┐
   9 │  □    │   □    │
     └───────┴────────┘
```

2. $22 \times 6 =$ _____

Name ....

Use the Distributive Property to multiply. Draw an area model.

3. 32
      × 7

4. 15
      × 8

5. 11
      × 8

6. 63
      × 4

7. 55
      × 6

8. 49
      × 9

Algebra Find each unknown number.

9. $37 \times 5 = s$

$s = $ _____

10. $99 \times 9 = t$

$t = $ _____

11. $85 \times 5 = v$

$v = $ _____

12. Write an equation that represents the area model below.

_____ × _____ = _____

Problem Solving

13. Mr. Kline bought 4 sheets of postage stamps. Each sheet had 16 stamps. What is the total number of postage stamps Mr. Kline bought? Write an equation to solve.

$s =$ _____ × _____ ,

$s =$ _____

Mr. Kline bought _____ stamps.

14. The World of Maps store displays its maps on 3 shelves. There are 26 world maps on each shelf. How many world maps does the store have to sell? Write an equation to solve.

$m =$ _____ × _____

$m =$ _____

The store has _____ world maps.

Brain Builders

15. **Processes &Practices** 3 **Find the Error** Candace is finding 67×2. She thinks the product is 124. Use the Distributive Property and an area model to explain and correct her mistake.

Draw an area model to support your answer.

_____

_____

16. **Building on the Essential Question** How can the Distributive Property help when you are multiplying by a two-digit number?

_____

_____

_____

Lesson 7

The Distributive Property

Homework Helper

Need help? connectED.mcgraw-hill.com

Find 8 × 16.

Think of 8 × 16 as (8 × 10) + (8 × 6).

Helpful Hint

Using the Distributive Property, you can multiply the addends of a number. Then add the products.

$8 \times 16 = (\mathbf{8 \times 10}) + (8 \times 6)$

$= \quad 80 \quad + \quad 48$

$= \quad\quad 128$

So, 8 × 16 = 128.

Practice

Use the Distributive Property to multiply. Draw an area model.

1. 28
    × 4

2. 19
    × 5

3. 41
    × 6

Multiply. Use the Distributive Property.

4. $75 \times 6 =$ _____   **5.** $4 \times 52 =$ _____   **6.** $8 \times 38 =$ _____

7. $97 \times 2 =$ _____   **8.** $7 \times 63 =$ _____   **9.** $6 \times 33 =$ _____

Brain Builders

10. **Processes &Practices** **3** **Justify Conclusions** Paula can make 14 beaded bracelets in an hour. Tina can make 13 woven bracelets in an hour. In one week, Paula made bracelets for 6 hours, and Tina made bracelets for 8 hours. Who made more bracelets that week? Explain.

_____

_____

11. Alejandro owns 3 sets of trains. Jess owns 2 sets of trains. Each set has 18 cars. How many train cars do they have in all?

_____

Vocabulary Check

Write *Distributive Property* or *decompose* on each line.

12. To _____ a number means to break it into parts.

13. The _____ states that you can multiply addends of a number, then add the products.

14. **Test Practice** Which expression represents the model?

Ⓐ $(5 \times 10) \times (5 \times 3)$

Ⓑ $5 \times 5 \times 5 \times 3$

Ⓒ $(5 \times 10) + 3$

Ⓓ $(5 \times 10) + (5 \times 3)$

Name _____

Base-ten blocks can be used to multiply two-digit numbers.

Math in My World

Tools Watch Tutor

Example 1

Zach bought 13 packages of lightbulbs. There are 4 lightbulbs in each package. How many lightbulbs are there in all?

Find 4×13.

1 Model 4 groups of 13.

2 Multiply the ones, and regroup.

$$\begin{array}{r} \square \\ 1\ 3 \\ \times \quad 4 \\ \hline \square\ \square \end{array}$$

10 { ... + 4 { ...

4   3

12 ones = 1 ten, 2 ones

3 Multiply the tens. Add the regrouped ten.

So, $4 \times 13 =$ _____.

4 groups of 10   1 ten   2 ones

Check Use the Distributive Property.

$4 \times 13 = (4 \times 10) + (4 \times 3)$

$= \underline{\quad} + \underline{\quad} = \underline{\quad}$

Example 2

Will makes $6 an hour shampooing dogs at a pet shop. Last month he worked 38 hours. How much money did Will earn?

Find 38 × $6.

Estimate   38 × 6 ⟶ _____ × _____ = _____

1 **Multiply the ones.**

   6 × 8 = 48

2 **Regroup 48 ones as:**

   ———— 4 tens and

   ———— 8 ones.

3️⃣ **Multiply the tens.**

   6 × 3 = 18

4 **Add the regrouped tens.**

   18
   + 4
   ———
   22

So, Will earned _____.

Check for Reasonableness

The product, _____, is close to the estimate, _____.

Guided Practice

Multiply. Check for reasonableness.

1.   23
        × 4

2.   42
        × 6

Explain how to find 6 × 37.

Estimate: Estimate:

_____ × _____ = _____   _____ × _____ = _____

Name _____

Multiply. Check for reasonableness.

3.   33
      × 5

Estimate: _____

4. $24
     × 4

Estimate: _____

5.   13
      × 7

Estimate: _____

6. 29 × 4 = _____

Estimate: _____

7. 5 × 18 = _____

Estimate: _____

8. 7 × $36 = _____

Estimate: _____

9. 6 × 52 = _____

Estimate: _____

10. 75 × 8 = _____

Estimate: _____

11. 4 × $83 = _____

Estimate: _____

Algebra Find the unknown number in each equation.

12. 5 × 31 = x

13. 63 × 7 = m

14. 49 × 8 = w

x = _____

m = _____

w = _____

Problem Solving

A local cave has walking tours. Adult tickets cost $18. Child tickets cost $15. Gemstone panning costs $12 per person.

15. The Diaz family has 2 adults and 3 children. How much would it cost for the family to go on a walking tour?

_____

16. **Processes &Practices** 2 **Reason** Can the Diaz family pan for gemstones for $75? Explain.

_____

_____

17. Find the total cost for the Diaz family to take the walking tour and pan for gemstones.

_____

_____

Brain Builders

18. **Processes &Practices** 2 **Use Number Sense** Write two multiplication problems that, when the products are added together, have a sum of 240.

_____

19. **Processes &Practices** 3 **Which One Doesn't Belong?** Circle the multiplication problem that does not belong with the other three. Explain.

$$\begin{array}{r} 12 \\ \times\ 8 \\ \hline \end{array} \qquad \begin{array}{r} 22 \\ \times\ 4 \\ \hline \end{array} \qquad \begin{array}{r} 42 \\ \times\ 2 \\ \hline \end{array} \qquad \begin{array}{r} 33 \\ \times\ 3 \\ \hline \end{array}$$

_____

20. **Building on the Essential Question** What steps can I use to multiply by a two-digit number with regrouping?

_____

Name _____

MY Homework

Homework Helper eHelp

Need help? 🖱 connectED.mcgraw-hill.com

Find 5 × 37.

Estimate 5 × 40 = 200

Multiply.

$$
\begin{array}{r}
{\scriptstyle 3} \\
3\,7 \\
\times \quad 5 \\
\hline
1\,8\,5
\end{array}
$$

1 Multiply the ones.
5 × 7 = 35
Regroup 35 ones as 3 tens and
5 ones. Write 5 in the ones place.
Write 3 above the tens place.

2 Multiply the tens.
5 × 3 = 15
Add the regrouped tens:
15 + 3 = 18
18 tens is 1 hundred and 8 tens.

Check for Reasonableness

The product, 185, is close to the estimate, 200.

You can also use the Distributive Property to check:

(5 × 30) + (5 × 7) = 150 + 35 = 185

Practice

Multiply. Check for reasonableness.

1. $\begin{array}{r} 77 \\ \times\ 3 \\ \hline \end{array}$

2. $\begin{array}{r} \$54 \\ \times\ 6 \\ \hline \end{array}$

3. $\begin{array}{r} 35 \\ \times\ 4 \\ \hline \end{array}$

Estimate: _____

Estimate: _____

Estimate: _____

Multiply. Check for reasonableness.

4. 8 × $46 = _____

5. 2 × 93 = _____

6. 7 × 68 = _____

Estimate: _____

Estimate: _____

Estimate: _____

7. 4 × 57 = _____

8. $7 × 13 = _____

9. 5 × $85 = _____

Estimate: _____

Estimate: _____

Estimate: _____

Brain Builders

10. Mrs. Sands teaches 4 classes at the high school and 2 at the community center. There are 36 students in each class. How many students does she teach in all?

_____

11. Vin charges $25 rent each day for a booth at his flea market. If 4 people rent a booth Saturday, and 4 people rent a booth Sunday, how much rent money will Vin collect?

_____

12. **Processes &Practices** **2** **Use Number Sense** Becky works 16 days in June and 14 days in July. If she drives 6 miles on the days she works, how many miles will she drive in June and July?

_____

13. **Test Practice** There are 25 gold paper clips and 75 silver paper clips in each box. How many silver paper clips are in 8 boxes?

Ⓐ 800

Ⓒ 600

Ⓑ 640

Ⓓ 200

Name ....

Lesson 9
Multiply by a Multi-Digit Number

ESSENTIAL QUESTION
How can I communicate multiplication?

You can use partial products to multiply by a multi-digit number.

 Math in My World Watch Tutor

Example 1

Today is Laura's birthday, and she is nine years old. Except for leap years, there are 365 days in one year. How many days old is Laura?

Find 365 × 9.

Estimate 9 × 365 ⟶ 9 × _____ = _____

$$
\begin{array}{r}
3\ 6\ 5 \\
\times \qquad 9 \\
\hline
\square\ \square \\
\square\ \square\ \square \\
+\ \square\ ,\ \square\ \square \\
\hline
\square\ ,\ \square\ \square
\end{array}
$$

Multiply 9 × 5.

Multiply 9 × 60.

Multiply 9 × 300.

Add the partial products.

	300	+ 60	+ 5
9	2,700	540	45

So, Laura is _____ days old.

Check for Reasonableness

The product, _____, is close to the estimate, _____.

Example 2 Tutor

Find 3 × $1,175.

$$
\begin{array}{r}
\boxed{}\boxed{} \\
\$1,\ 1\ 7\ 5 \\
\times \qquad\quad 3 \\
\hline
\$\ \boxed{}\ ,\boxed{}\boxed{}\boxed{}
\end{array}
$$

1 Multiply the ones.

3 × 5 ones = 15 ones

Regroup 15 ones as 1 ten and 5 ones.

2 Multiply the tens.

3 × 7 tens = 21 tens

Add the regrouped tens.

21 tens + 1 ten = 22 tens

Regroup 22 tens as 2 hundreds

and 2 tens.

3 Multiply the hundreds.

3 × 1 hundred = 3 hundreds

Add the regrouped hundreds.

3 hundreds + 2 hundreds = 5 hundreds

4 Multiply the thousands.

3 × 1 thousand = 3 thousands

Guided Practice

Multiply. Check for reasonableness.

1. 135
    × 2

2. 532
    × 6

Talk MATH

Explain why it is a good idea to estimate answers to multiplication problems.

Independent Practice

Multiply. Check for reasonableness.

3. 313
     × 3

4. 819
     × 5

5. $791
     ×   5

6. 238
     × 4

7. 7 × $460 = _____

8. 7 × 561 = _____

9. 8 × 6,328 = _____

10. 9 × $5,679 = _____

Algebra Find each unknown number.

11. 8 × 7,338 = x

12. 7 × 8,469 = y

13. 9 × $9,927 = t

14. 9 × 8,586 = u

x = _____

y = _____

t = _____

u = _____

Algebra Find each product if $n = 8$.

15. n × 295 = _____

16. 737 × n = _____

17. n × $2,735 = _____

Compare. Use >, <, or =.

18. 4 × 198 ◯ 3 × 248

19. 7 × 385 ◯ 6 × 457

Problem Solving

20. Mr. Gibbons buys 8 cases of seeds at the school plant sale. If there are 144 packages of seeds in each case, how many packages of seeds has he bought?

_____

21. On average, 1,668 gallons of water are used daily by each person in the United States. How much water is used by one person in a week?

_____

22. Each set of furniture costs $2,419. How much would it cost to buy 5 sets of furniture?

_____

 Brain Builders

Processes
23. &Practices ▷1▷ **Keep Trying** Complete the equation.

$$\boxed{}, 287 \times 6 = 25, \boxed{} 2 \boxed{}$$

Processes
24. &Practices ▷7▷ **Identify Structure** Write a four-digit number and a one-digit number whose product is greater than 6,000 and less than 6,200. Explain your steps.

_____

_____

25. **Building on the Essential Question** How is multiplying by multi-digit numbers similar to multiplying by two-digit numbers?

_____

_____

_____

Name

MY Homework

Homework Helper

Need help? connectED.mcgraw-hill.com

Find 3 × 2,763.

Estimate the product: 3 × 3,000 = 9,000

$$\begin{array}{r} 2\ \ 1\ \ \ \\ 2,7\ 6\ 3 \\ \times\ \ \ \ \ \ 3 \\ \hline 8,2\ 8\ 9 \end{array}$$

1 Multiply the ones.

3 × 3 = 9

Write 9 in the ones place.

2 Multiply the tens.

3 × 6 = 18

Regroup 18 tens as 1 hundred and 8 tens.

Write 8 in the tens place.

3 Multiply the hundreds.

3 × 7 = 21

Add the regrouped ten.

21 + 1 = 22

Regroup 22 hundreds as 2 thousands
and 2 hundreds.

Write 2 in the hundreds place.

4 Multiply the thousands.

3 × 2 = 6

Add the regrouped hundreds.

6 + 2 = 8

Write 8 in the thousands place.

Check for Reasonableness

The product, 8,289, is close to the estimate, 9,000.

Another way to check is using partial products.

Multiply 3 × 3.
Multiply 3 × 60.
Multiply 3 × 700.
Multiply 3 × 2,000.
Add the partial products.

6,000   + 2,100 + 180 + 9 = 8,289

Multiply. Check for reasonableness.

1. 1,313
   × 9

2. $547
   × 6

3. 6,421
   × 3

4. $4,512
   × 5

5. 3,525 × 6 = _____

6. 7 × 7,441 = _____

Algebra Find each product.

7. $n = 8$
   $n × \$685 =$ _____

8. $n = 3$
   $n × 5,266 =$ _____

Brain Builders

9. **Processes &Practices** 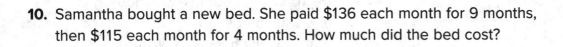 **Use Number Sense** One shelf in the greenhouse holds 367 pepper plants and 100 tomato plants. How many plants can 6 shelves hold?

   _____

10. Samantha bought a new bed. She paid $136 each month for 9 months, then $115 each month for 4 months. How much did the bed cost?

    _____

11. A concert hall seats 802 people in the balcony and 6,887 on the main floor. There were 8 concerts in June, and a ticket was sold for every seat. How many tickets were sold in June?

    _____

12. **Test Practice** Find the product $n × 2,019$ if $n = 5$.

    Ⓐ 10,000   Ⓒ 10,095

    Ⓑ 10,055   Ⓓ 10,545

Check My Progress

Vocabulary Check

1. Circle the example that correctly shows how to use the **Distributive Property** to find the product of 5 × 15.

$(5 \times 10) + (5 \times 5)$

$(5 \times 10) \times (5 \times 5)$

$(5 \times 10) \times (5 \times 15)$

$(5 \times 10) + (5 \times 15)$

2. Explain how to use **partial products** to multiply.

_____

_____

3. When you use place value to exchange equal amounts when renaming a number, what are you doing?

_____

Concept Check

Multiply. Check for reasonableness.

4.
$$\begin{array}{r} 23 \\ \times\ 2 \\ \hline \end{array}$$

5.
$$\begin{array}{r} 227 \\ \times\ 8 \\ \hline \end{array}$$

6.
$$\begin{array}{r} 45 \\ \times\ 7 \\ \hline \end{array}$$

7.
$$\begin{array}{r} 612 \\ \times\ 4 \\ \hline \end{array}$$

Problem Solving

8. The table shows prices of items at an electronics store.

Electronics Store	
Item	**Price**
Battery pack	$13
Copper wire	$22

How much would it cost to buy 3 battery packs and 3 copper wires?

_____

9. Each campsite needs the number of lanterns shown below. How many lanterns are needed for 48 campsites?

Brain Builders

10. There are 1,440 minutes in a day. How many minutes are in 7 days? Explain how to find how many minutes are in 2 weeks.

_____

11. Test Practice Mohammed used an area model to show 4 × 35.

What is the missing number?

Ⓐ 3   Ⓒ 30

Ⓑ 5   Ⓓ 50

Name

ESSENTIAL QUESTION
How can I communicate multiplication?

Learn the Strategy

Tutor

There are five girls in Dina's scout troop. They are going to an amusement park. Children's tickets cost $22. What is the total cost of admission? Is an estimate or an exact answer needed?

1 Understand

What facts do you know?

The price of admission is _____ for each girl. There are _____ girls.

What do you need to find?

the total cost of admission and whether an estimate or exact answer is needed

2 Plan

The girls will need to know exactly how much to pay.

So, they need an _____ answer. Find _____ × _____.

3 Solve

$$\begin{array}{r} 22 \\ \times\ 5 \\ \hline \end{array}$$

So, the troop needs _____ to go to the amusement park.

4 Check

Does your answer make sense? Explain.

_____

Online Content at **connectED.mcgraw-hill.com**

Each motorized scooter at the scooter store costs $75. About how much would 7 motorized scooters cost? Is an estimate or an exact answer needed?

 Understand

What facts do you know?

_____

_____

What do you need to find?

_____

 Plan

_____

_____

Solve

 Check

Does my answer make sense? Explain.

_____

Name
....

Apply the Strategy

Determine whether each problem requires an estimate or an exact answer. Then solve.

1. An office needs to buy 6 computers and 6 printers. Each computer costs $384. Each printer costs $88. About $2,400 will be spent on computers. What is the question?

   _____

   _____

   _____

2. Each fourth grade class reads a total of 495 minutes each week. Suppose there are 4 fourth grade classes. How many minutes are the 4 classes reading each week?

   _____

   _____

   _____

Brain Builders

3. There are 12 stickers on each sheet. There are three sheets in one pack. Delinda wants 100 stickers. How many packs should she buy?

   _____

   _____

4. **Processes &Practices** **3** **Draw a Conclusion** Determine if Tammy, Anessa, and Jaleesa have more than 110 CDs.

   _____

   _____

Name	CDs Owned
Tammy	21
Anessa	42
Jaleesa	33

Review Strategies

Use any strategy to solve each problem.

- Use the four-step plan.
- Check for reasonableness.
- Estimate or find an exact answer.

5. Each day, Sparky eats the number of treats shown. How many treats does Sparky eat in one year? (*Hint:* There are 365 days in a year.)

Sparky eats _____ treats in one year.

6. **Processes &Practices**  **Find the Error** Each fourth grade class at Tannon Elementary School wants to raise $475 for a local charity. There are 5 fourth grade classes. Rianna says that the overall goal is $2,055. Find and correct her mistake.

_____

_____

_____

7. An electronics store sells remote control dinosaurs for $395 each. About how much do four remote control dinosaurs cost?

_____

_____

8. There are 63 runners in a race. Each runner pays the amount shown to run. How much do the runners pay in all?

63 × $7 = _____

MY Homework

Homework Helper

Need help? connectED.mcgraw-hill.com

Tyrone wants to hand out fliers about a sale to each business in his neighborhood. He plans to leave 3 fliers at each shop, and there are 38 shops. About how many fliers will Tyrone need? Is an estimate or exact answer needed?

1 Understand

What do you know?

Tyrone plans to leave 3 fliers at each shop. There are 38 shops.

What do you need to find?

I need to find about how many fliers Tyrone needs.

2 Plan

The question asks *about* how many fliers are needed. The word *about* means an exact answer is not necessary. The question is asking for an estimate. I will round 38 and multiply.

3 Solve

$$
\begin{array}{ccc}
38 & \text{rounds to} & 40 \\
\times\ 3 & & \times\ 3 \\
\hline
& & 120
\end{array}
$$

So, Tyrone needs about 120 fliers.

4 Check

I will check by comparing the exact answer to the estimate.

$38 \times 3 = (30 \times 3) + (8 \times 3) = 114$

114 is close to the estimate, 120, so the answer is reasonable.

Problem Solving

Determine whether each problem requires an estimate or exact answer. Then solve.

1. Jeff is having a dinner party. He has a large rectangular table that seats 10 people on each long side and 4 people on each of the two ends. How many people can sit at Jeff's table?

2. Brittany borrowed 3 movies from the library. Each movie is almost 2 hours long. About how many hours of movies does Brittany have to watch?

3. Matt is running the family fun fair at school. He collected about $65 in donations each month for the 7 months he was planning the fair. About how much money does Matt have to spend for the fair?

Brain Builders

4. Fatima works at a bakery. She places 5 candied flowers on top of each cupcake she decorates. She will decorate 2 dozen cupcakes today and 2 dozen tomorrow. How many candied flowers will Fatima use in these two days? (*Hint:* 1 dozen = 12)

5. It takes Kayla 12 minutes to walk from her house to her grandparents' house. She walks to and from their house twice a week. About how many minutes each week does Kayla spend walking between the two houses?

6. **Processes &Practices** 2 **Use Number Sense** A haiku is a poem with exactly 3 lines. If each of the 26 students in Mr. Kopp's class write two haiku, how many lines of poetry does the class write in all?

Name ....

ESSENTIAL QUESTION
How can I communicate multiplication?

You can use the Distributive Property or partial products to multiply across zeros.

  Math in My World Watch ▶ Tutor 💬

Example 1

Iván's braces cost $108 each month. How much will Iván's parents pay in 6 months?

Find 6 × $108.

Estimate 6 × $108 ⟶ 6 × _____ = _____

	$100 +	$8
6	6 × $100	6 × $8

← 6 × 0 = 0, so there is no space in the rectangle for that product.

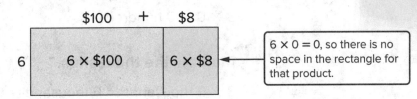

One Way **Distributive Property**

6 × $108 = (6 × $100) + (6 × $8)

= (☐) + (☐)

= ☐

Another Way **Partial Products**

$$\begin{array}{r} \$108 \\ \times\ \ 6 \end{array}$$

$ ☐ ☐   6 × $8

$ ☐   6 × $0

+ $ ☐ ☐ ☐   6 × $100
_____

$ ☐ ☐ ☐   Add the partial products.

So, Iván's parents will pay _____ in 6 months.

Check for Reasonableness

The answer, _____, is close to the estimate, _____.

Example 2

If three trees are each 2,025 years old, what is the total age of the trees?

Estimate 3 × 2,025 ⟶ _____ × _____ = _____

2,0 2 5
×   3
_____

1 **Multiply the ones.**

3 × 5 ones = 15 ones
Regroup 15 ones as 1 ten
and 5 ones.

2 **Multiply the tens.**

3 × 2 tens = 6 tens
Add the regrouped tens.
6 tens + 1 ten = 7 tens

3 **Multiply the hundreds.**

3 × 0 = 0 hundreds

4 **Multiply the thousands.**

3 × 2 thousands = 6 thousands

So, the total age of the trees is _____ years.

Check for Reasonableness

The answer, _____, is close to the estimate, _____.

Guided Practice

Multiply. Check for reasonableness.

1.   303
    ×   3

2. $507
     ×   6

Estimate:

3 × 300 = _____

Estimate:

6 × $500 = _____

Explain how to find the product of 4 and 2,008.

Name _____

Multiply. Check for reasonableness.

3.   201
   × 2

4. $402
   × 3

5.   709
   × 5

Estimate:

_____ × _____ = _____

Estimate:

_____ × _____ = _____

Estimate:

_____ × _____ = _____

6. 904 × 9 = _____

7. 2 × $1,108 = _____

8. 4 × 6,037 = _____

Estimate:

Estimate:

Estimate:

9. 8,504 × 3 = _____

10. 6 × 6,007 = _____

11. 5 × $9,082 = _____

Estimate:

Estimate:

Estimate:

Algebra Find the unknown number.

12. $6 \times 4,005 = s$

13. $9,002 \times 9 = q$

14. $8,009 \times 7 = r$

$s =$ _____

$q =$ _____

$r =$ _____

Problem Solving

15. **Processes &Practices** **Stop and Reflect** Large pool equipment sets cost $1,042. Small pool equipment sets cost $907. How much does it cost to buy 3 large pool equipment sets?

How much more does it cost to buy 2 large pool equipment sets than 2 small pool equipment sets?

16. Diller Elementary is collecting money to donate to a charity. $103 is collected each month. How much money is collected in 9 months?

Brain Builders

17. **Processes &Practices** **Understand Symbols** Complete the number sentence. Explain what the symbols represent.

☐,☐☐☐ × ☐ = ☐☐,☐☐☐

18. **Processes &Practices** 3 **Which One Doesn't Belong?**
Circle the expression that does not belong. Explain.

4,006 × 5   3,015 × 2

2,010 × 3   1,206 × 5

19. **Building on the Essential Question** Why do the products of multi-digit numbers with zeros and one-digit numbers sometimes have zeros in them and sometimes have no zeros in them?

Name ....

MY Homework

Homework Helper eHelp

Need help? ⟋ connectED.mcgraw-hill.com

Find 4 × 1,405.

Estimate the product. 4 × 1,000 = 4,000

Multiply.

```
  1   2
1,405
×   4
5,620
```

So, 4 × 1,405 = 5,620.

 Multiply the ones.
4 × 5 = 20
Regroup 20 ones as
2 tens and 0 ones.
Write 0 in the ones place.

Multiply the tens.
4 × 0 = 0
Add the regrouped ones.
0 + 2 = 2
Write 2 in the tens place.

Multiply the hundreds.
4 × 4 = 16
Regroup 16 hundreds as
1 thousand and 6 hundreds.
Write 6 in the hundreds place.

Multiply the thousands.
4 × 1 = 4
Add the regrouped hundreds.
4 + 1 = 5
Write 5 in the thousands place.

Check for Reasonableness The product, 5,620, is close to the estimate, 4,000.

Practice

Multiply. Check for reasonableness.

1.
```
 709
×  6
```

2.
```
 905
×  5
```

3.
```
5,079
×   8
```

4.
```
2,006
×   4
```

Multiply. Check for reasonableness.

5. 5,001
    ×   9

6. 4,807
    ×   7

7. 3,004
    ×   3

8. 8,060
    ×   3

9. 6,010 × 8 = _____

10. 9,012 × 6 = _____

11. 2 × 1,805 = _____

12. 4 × 1,009 = _____

Brain Builders

13. The art teacher ordered 101 boxes of markers and 101 boxes of chalk for his classroom. If each box contains 8 items, how many items did he order in all? Show your work.

14. Brent rode his bicycle 4 miles on Saturday. There are 1,760 yards in a mile. How many yards did Brent ride on Saturday? Explain what the zeros in your answer mean.

15. **Processes &Practices** **2** **Use Number Sense** Lorna packs 1,024 socks in each large box and 448 socks in each small box at the sock factory. What is the total number of socks in 7 large boxes? What is the total number of socks in 4 small boxes?

16. **Test Practice** There are 405 windows in the office building. Each window has 9 panes. How many window panes does the building have in all?

   Ⓐ 4,059   Ⓑ 3,645   Ⓒ 3,600   Ⓓ 4,009

Review

Vocabulary Check

Choose the correct word(s) to complete each sentence.
Find the word(s) in the puzzle.

S	T	C	U	D	O	R	P	L	A	I	T	R	A	P	N	F	G	P	D
Y	J	G	S	U	L	X	W	D	M	M	A	R	M	A	A	N	N	R	G
D	P	A	X	Q	T	Y	R	T	Y	U	U	R	P	C	N	T	T	O	P
F	K	E	D	M	T	M	E	N	S	K	R	L	T	U	H	X	P	D	P
F	S	Z	S	T	K	M	I	X	N	I	B	O	T	M	I	S	U	U	N
Q	A	W	D	J	L	O	C	K	O	Z	R	V	V	I	T	B	O	C	O
D	I	S	T	R	I	B	U	T	I	V	E	P	R	O	P	E	R	T	Y
A	H	E	K	C	H	N	K	C	P	A	C	I	P	E	L	L	G	T	C
K	I	H	L	B	K	W	J	A	E	C	R	T	S	X	K	L	E	J	I
W	A	N	A	E	S	T	I	M	A	T	E	B	U	A	Y	P	R	P	E

Distributive Property

estimate

factor

multiple

partial products

product

regroup

1. The answer or result of a multiplication problem is the _____.

2. A(n) _____ is a number close to an exact value.

3. A(n) _____ of a number is the product of that number and any whole number.

4. To _____ is to use place value to exchange equal amounts when renaming a number.

5. The _____ combines multiplication and addition by multiplying each addend by the number and adding the products.

6. To use _____, find the product for the ones, tens, and so forth separately, and then add them together.

7. A number that divides a whole number evenly is called a _____. It is also a number that is multiplied by another number.

Concept Check

Multiply. Use basic facts and patterns.

8. $4 \times 90 =$ _____

9. $6 \times 3,000 =$ _____

Estimate. Round to the greatest place value. Circle if the estimate is _greater than_ or _less than_ the actual product.

10. $1,478 \times 4$

_____ $\times 4 =$ _____

greater than

less than

11. $5 \times 6,225$

$\downarrow$

$5 \times$ _____ $=$ _____

greater than

less than

Multiply. Check for reasonableness.

12. $\begin{array}{r} 43 \\ \times\ 5 \\ \hline \end{array}$

13. $\begin{array}{r} 24 \\ \times\ 9 \\ \hline \end{array}$

14. $\begin{array}{r} 829 \\ \times\ \ 8 \\ \hline \end{array}$

15. $\begin{array}{r} 724 \\ \times\ \ 7 \\ \hline \end{array}$

16. $\begin{array}{r} 569 \\ \times\ \ 6 \\ \hline \end{array}$

17. $\begin{array}{r} 617 \\ \times\ \ 3 \\ \hline \end{array}$

Problem Solving

18. Marcia read 2 books. Each book was 44 pages long. How many pages did she read?

_____

19. Tania has four decks of 52 cards in each deck. How many cards does Tania have?

_____

20. Each ski pass costs $109. How much do five ski passes cost?

_____

Brain Builders

21. A kangaroo can jump as far as 44 feet in a single jump. What distance would three jumps of this size cover? Explain how to use the Distributive Property to solve.

_____

44 feet

22. There are 4 blue party bags and 4 red party bags. Each bag contains 12 items. Is it reasonable to say that the bags will have 75 items in all? Explain.

_____

23. **Test Practice** A plane carries 234 passengers. If the plane makes 4 trips a day, how many passengers does the plane transport a day?

  Ⓐ 51 passengers   Ⓒ 826 passengers

  Ⓑ 800 passengers   Ⓓ 936 passengers

Use what you learned about multiplication
to complete the graphic organizer.

ESSENTIAL QUESTION

How can I communicate
multiplication?

Models

**Partial
Products**

**Distributive
Property**

Example

Reflect on the ESSENTIAL QUESTION Write your answer below.

Performance Task

Dude Ranch

This summer, Eli is spending a month at a dude ranch that is 513 miles from his home. He is learning about ranching, caring for the ranch animals, and horseback riding.

Show all your work to receive full credit.

Part A

Eli's parents drove him to the ranch in the family car, and then returned home. They will drive the same route to pick Eli up at the end of the month. Eli's older sister Becca plans to drive the family car to the ranch in the middle of the month to visit him, and then return home. How many total miles will the car be driven to and from the ranch this summer? Show your work in the space at the right.

Part B

Becca says that she can usually drive 58 miles per hour on the highway that leads to the ranch. Is it reasonable to think that Becca can make it to the ranch in 7 hours if she drives at that rate? Explain your answer using estimation.

Part C

Eli was inspired to go to a dude ranch when he learned about the Mongol Derby, the world's longest horse race. The horses race about 620 miles. The Mongol Derby has been run every year for 8 years. Explain how you can use multiplication of a one-digit number by a two-digit number to find about how many miles the race has covered in 8 years.

Part D

Part of Eli's job on the ranch is to feed the new calves. There are 5 new calves, and each calf eats 15 pound of hay every day. Complete the area model to show the total number of pounds of hay Eli feeds the calves in 6 days. Then write the partial products and the total product.

Part E

Eli is going on a three-day ride with 13 other dudes, 3 horse wranglers, and 1 trail boss. Each person and horse team needs 84 pounds of supplies for the trip. The ranch has 5 wagons, but they only want to haul one wagon for the group. Use estimation to figure out which wagons can haul all the needed supplies for the ride. Explain your answer.

Wagon	A	B	C	D	E
Maximum Weight (pounds)	1,500	2,200	1,800	1,200	1,750

5 Multiply with Two-Digit Numbers

ESSENTIAL QUESTION

How can I multiply by a two-digit number?

Animals in MY world

Watch a video!

Watch

MY Chapter Project

Name That Digit!

1. Write a multiplication problem that multiplies by a 2-digit number with its answer on the card provided for you at the bottom corner of this page. After you have written and checked your answer, cut out the card and give it to your teacher.

2. You will be a contestant on a game show! Your teacher will assign your class to teams.

3. Your teacher will shuffle all of the cards, and write one on the board, but it will be missing one of its digits. As a team, you have 30 seconds to discuss with your group to find the missing digit and write it on a piece of paper.

4. After 30 seconds, your teacher will check your answers. If you get it correct, your team receives one point. The first team to earn 15 points wins the game.

X _____

_____

Name ....

Round to the given place.

1. 85,888; nearest ten thousand

_____

2. 681,002; nearest hundred thousand

_____

3. The students raised $6,784 for a new playground. To the nearest thousand, about how much money did the students raise?

_____

Add.

4.
```
   759
+  307
```

5.
```
  34,068
+  6,055
```

6.
```
  242,607
+ 480,196
```

Write a multiplication equation that represents each model.

7.

8.

_____ × _____ = _____   _____ × _____ = _____

Multiply.

9. 40 × 9 = _____

10. 36 × 7 = _____

Name

Review Vocabulary

decompose   equation   factor   product

Making Connections

Use the review vocabulary to complete each section of the bubble map.

Write a sentence about or an example for each word.

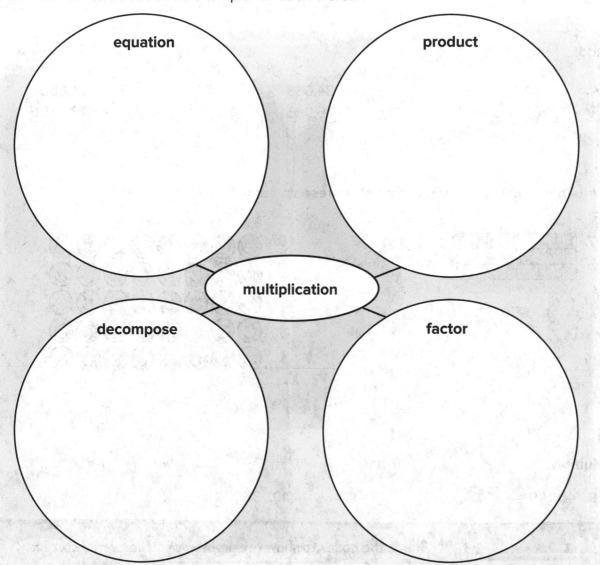

equation

product

multiplication

decompose

factor

MY Vocabulary Cards

Processes
&Practices

Lesson 5–5

operation

$$+ \quad | \quad -$$
$$\times \quad | \quad \div$$

Ideas for Use

- Write a tally mark on each card every time you read the word in this chapter or use it in your writing. Challenge yourself to use at least 10 tally marks for each card.

- Use the blank cards to write your own vocabulary cards.

A mathematical process such as addition, subtraction, multiplication, or division.

Which operations would you use in these equations?

$$200 \bigcirc 4 = 800$$

$$874 \bigcirc 555 = 319$$

MY Foldable

FOLDABLES® Follow the steps on the back to make your Foldable.

ESTIMATE

PRODUCTS

Problem

$$23 \times 41$$

Estimate

Solve

Reasonable?

yes

no

....

Lesson 1
Multiply by Tens

ESSENTIAL QUESTION
How can I multiply by a
two-digit number?

 Math in My World Watch ▶ Tutor

Example 1

**Miss Rita took 20 pictures at the zoo.
She printed the pictures so that each of
her 25 students could have a copy. How
many pictures did Miss Rita print?**

Find 25×20.

The number 20 is a multiple of ten.

One Way Use properties.

Think of 20 as _____ × 10.

$25 \times 20 = 25 \times (\underline{\hspace{1cm}} \times \underline{\hspace{1cm}})$

$ = (25 \times 2) \times 10$

$ = 50 \times 10 = \underline{\hspace{2cm}}$

You have used the Associative Property
of Multiplication.

Another Way Use paper and pencil.

1 Multiply the ones.

$$\begin{array}{r} 2\ 5 \\ \times\ 2\ 0 \\ \hline \end{array}$$

☐ ← ⎡ 0 ones × 25 = ⎤
⎣ _____ ⎦

2 Multiply the tens.

$$\begin{array}{r} 2\ 5 \\ \times\ 2\ 0 \\ \hline \end{array}$$

☐ ☐ ☐ ← ⎡ 2 tens × 25 = ⎤
⎣ _____ tens ⎦

So, Miss Rita printed _____ pictures.

Online Content at 🖱 **connectED.mcgraw-hill.com**

Example 2

An electronics store has 30 digital music players in stock that cost
$99 each. How much do the digital music players cost all together?

You need to find $99 × 30. The number 30 is a multiple of 10.

1 **Multiply the ones.**

0 ones × 99 = _____

$$\begin{array}{r} \$9\ \ 9 \\ \times\ \ 3\ \ 0 \\ \hline \boxed{} \end{array}$$

2 **Multiply the tens.**

3 tens × 99

= _____ tens

$$\begin{array}{r} \$9\ \ 9 \\ \times\ \ \underline{\hphantom{99}\ 3\ \ 0} \\ \boxed{}\,,\boxed{}\boxed{}\boxed{} \end{array}$$

> **Helpful Hint**
> When you multiply a
> number by a multiple of 10,
> the digit in the ones place
> is always zero.

So, the music players cost a total of $ _____.

Guided Practice

Multiply.

1. $\begin{array}{r} 36 \\ \times\ 10 \\ \hline \end{array}$

2. $\begin{array}{r} 53 \\ \times\ 30 \\ \hline \end{array}$

3. $\begin{array}{r} 42 \\ \times\ 20 \\ \hline \end{array}$

4. $\begin{array}{r} 64 \\ \times\ 40 \\ \hline \end{array}$

Joey is finding 67 × 40.
Explain why he can think of
67 × 40 as 67 × 4 × 10.

Name ....

Multiply.

5. 15
    × 20

6. 27
    × 30

7. 46
    × 40

8. $53 \times 60 = $ _____

9. $80 \times 80 = $ _____

10. $94 \times 90 = $ _____

11. $\$27 \times 10 = $ _____

12. $\$31 \times 30 = $ _____

13. $\$38 \times 50 = $ _____

14. $\$45 \times 50 = $ _____

15. $\$56 \times 70 = $ _____

16. $\$69 \times 80 = $ _____

17. If $7 \times 29 = 203$, then what is 70×29?

_____

18. If $3 \times 52 = 156$, then what is 30×52?

_____

Algebra Use mental math to find the unknown number.

19. $22 \times y = 440$

$y = $ _____

20. $15 \times y = 450$

$y = $ _____

21. $25 \times z = 500$

$z = $ _____

Problem Solving

Hummingbirds feed every 10 minutes. They fly about 25 miles per hour and flap their wings 60 to 80 times each second.

22. What is the least number of times a hummingbird will flap its wings in 15 seconds?

_____

23. What is the greatest number of times a hummingbird will flap its wings in 15 seconds?

_____

24. How many minutes have passed if a hummingbird has eaten 45 times?

_____

Brain Builders

25. **Processes &Practices** 4 **Model Math** If a hummingbird flies 4 hours a day for 5 days, about how far did it fly? Write a number sentence to describe your answer.

_____

26. **Processes &Practices** 3 **Which One Doesn't Belong?** Circle the multiplication problem that does not belong with the other three. Rewrite the problem so that it does belong with the others. Explain your reasoning.

15 × 30	28 × 20	41 × 21	67 × 40

27. ❓ **Building on the Essential Question** How can place value help me multiply a two-digit number by a multiple of ten? Provide a worked out example that explains your reasoning.

_____

Name _____

MY Homework

Homework Helper

Need help? connectED.mcgraw-hill.com

Find 63×20.

1 Multiply the ones.

$$
\begin{array}{r}
63 \\
\times\ 20 \\
\hline
0
\end{array}
$$

0 ones × 63 = 0 →

2 Multiply the tens.

$$
\begin{array}{r}
63 \\
\times\ 20 \\
\hline
1{,}260
\end{array}
$$

2 tens × 63 = 126 tens →

Practice

Multiply.

1. $\begin{array}{r} 51 \\ \times\ 30 \\ \hline \end{array}$

2. $\begin{array}{r} 39 \\ \times\ 80 \\ \hline \end{array}$

3. $\begin{array}{r} 25 \\ \times\ 60 \\ \hline \end{array}$

4. $\begin{array}{r} 42 \\ \times\ 50 \\ \hline \end{array}$

5. $\begin{array}{r} 45 \\ \times\ 90 \\ \hline \end{array}$

6. $\begin{array}{r} 88 \\ \times\ 30 \\ \hline \end{array}$

7. $68 \times 40 =$ _____

8. $11 \times 70 =$ _____

9. $99 \times 10 =$ _____

Problem Solving

10. There are 40 rows of lockers. There are 12 lockers in each row. How many lockers are there?

_____

11. Pablo found out that every classroom has 34 desks. There are 30 classrooms. How many desks are in the school?

_____

12. It costs $10 per person to enter the museum. How much money would it cost for 30 people to enter the museum?

_____

Brain Builders

13. **Processes &Practices** **6** **Be Precise** Explain how the Commutative Property helps to find the unknown number in the equation $35 \times 70 = m \times 35$. Then solve for m.

_____

_____

14. **Algebra** Explain how to use mental math to find the unknown number in the equation $12 \times b = 480$. Solve.

_____

_____

15. **Test Practice** To raise money for an animal shelter, 17 students ran in a race. Each student raised $30. How much did the students raise in all?

Ⓐ $47 Ⓑ $51 Ⓒ $310 Ⓓ $510

Name _____

Lesson 2
Estimate Products

ESSENTIAL QUESTION
How can I multiply by a two-digit number?

The word *about* tells you to estimate. When you estimate the product of two 2-digit factors, it is helpful to round them both.

Math in My World

Watch ▶   Tutor 💬

Example 1

A hamster sleeps 14 hours each day. About how many hours does a hamster sleep in 3 weeks?
There are 21 days in 3 weeks.

So, estimate _____ × _____ .

1 **Round to the nearest ten.**

```
   21  ⟶  ☐ ☐
×  14  ⟶        21 rounded to the nearest 10 is _____.
           ×    14 rounded to the nearest 10 is _____.
       ⟶  ☐ ☐ ☐
```

2 Multiply.

So, a hamster sleeps about _____ hours in _____ days, or 3 weeks.

Since both factors were rounded down, the estimate is less than the actual product.

        21
    ┌─────────────┐
  14│Actual Product│
    └─────────────┘

        20   1
    ┌──────────┬──┐
 10 │ Estimate │ │
    └──────────┴──┘
  4

Online Content at 🔗 **connectED.mcgraw-hill.com**

Lesson 2 285

Example 2

Tonya spends 35 minutes playing at the park each day. About how many minutes does she play at the park in 38 days?

You need to estimate 38 × _____.

 Round each factor to the nearest ten.

38 rounded to the nearest 10 is _____.

35 rounded to the nearest 10 is _____.

 Multiply.

So, Tonya spends about _____ minutes playing at the park.

Since both factors were rounded up, the estimate is _____ than the actual product.

Guided Practice

1. **Estimate. Circle whether the estimate is *greater than* or *less than* the actual product.**

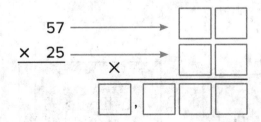

greater than

less than

Talk MATH

Explain how you know if an estimated product is greater than or less than the actual product.

Name ....

Estimate. Circle whether the estimate is *greater than* or *less than* the actual product.

2. 28 $\longrightarrow$
  $\times\ 25$ $\longrightarrow$ × ____

greater than

less than

3. 43 $\longrightarrow$
  $\times\ 14$ $\longrightarrow$ × ____

greater than

less than

4. $56 $\longrightarrow$
  $\times\ 37$ $\longrightarrow$ × ____

greater than

less than

5. 79 $\longrightarrow$
  $\times\ 55$ $\longrightarrow$ × ____

greater than

less than

6. $91 $\longrightarrow$
  $\times\ 64$ $\longrightarrow$ × ____

greater than

less than

7. 94 $\longrightarrow$
  $\times\ 82$ $\longrightarrow$ × ____

greater than

less than

Estimate the product.

8. $23 \times 11 =$ _____

9. $35 \times 37 =$ _____

10. $48 \times 86 =$ _____

11. $53 \times 42 =$ _____

12. $67 \times 56 =$ _____

13. $73 \times 84 =$ _____

Algebra Use mental math to find the unknown number.

14. $20 \times a = 1{,}200$

$a =$ _____

15. $b \times 30 = 900$

$b =$ _____

16. $40 \times c = 2{,}400$

$c =$ _____

Problem Solving

Use the information in the table for Exercises 17–18.
Write an equation to solve.

Green Darner Dragonfly Facts	
Average adult length	74 millimeters
Maximum length of nymph	55 millimeters

17. **Processes &Practices** 4 **Model Math** Suppose 18 nymph dragonflies of maximum length are laid end to end. About how long would they span?

    _____ × _____ = _____ mm

18. If 32 average adult dragonflies were laid end to end, about how long would they span?

    _____ × _____ = _____ mm

Algebra Write an equation to solve.

19. The art room has 15 shelves of paint. Each shelf has 48 cans of paint. About how many cans of paint are there all together?

    _____ × _____ = _____ cans of paint

20. There are 12 millipedes that each measure 16 centimeters long. About how long would they measure if they were laid end to end?

    _____ × _____ = _____ cm

Brain Builders

21. **Processes &Practices** 1 **Make a Plan** Identify two pairs of factors that have an estimated product of 2,000 so that one pair is an overestimate and one pair is an underestimate.

    _____

22. **Building on the Essential Question** How does an estimated product relate to the actual product? Explain.

    _____

    _____

    _____

Name ....

MY Homework

Homework Helper eHelp

Need help? connectED.mcgraw-hill.com

Estimate 88 × 65. Tell whether the estimate is _greater than_ or _less than_ the actual product.

1 Round each factor to the nearest ten.

$$88 \longrightarrow 90$$
$$\underline{\times\ 65} \longrightarrow \underline{\times\ 70}$$

88 rounds to 90.

65 rounds to 70.

2 Multiply.

$$
\begin{array}{r}
90 \\
\underline{\times\ 70} \\
6{,}300
\end{array}
$$

0 ones × 90 = 0

7 tens × 90 = 630 tens

The estimate for 88 × 65 is 6,300.

Since both factors were rounded up, the estimate is greater than the actual product.

Practice

Estimate.

1. 37 × 22 = _____ × _____ = _____

2. 87 × 41 = _____ × _____ = _____

3. 49 × 16 = _____ × _____ = _____

4. 25 × 12 = _____ × _____ = _____

Problem Solving

Estimate. Tell whether the estimate is *greater than* or *less than* the actual product.

Processes &Practices 4

5. **Model Math** A concert ticket costs $23. About how much will tickets cost for a group of 22 people?

_____

6. There are 32 times students can work in the computer lab during one week. If 24 students can work in the computer lab at one time, about how many students can work in the computer lab during one week?

_____

Brain Builders

Algebra **Write an equation to solve.**

7. Ramona creates 16 paintings each month. About how many paintings will she create in 3 years? Explain your reasoning.

_____ × _____ = _____ paintings

_____

_____

8. Michael averages 12 points in each basketball game. His team plays 2 games each week. About how many points will he score in 12 weeks? Explain your reasoning.

_____ × _____ = _____ points

_____

_____

9. **Test Practice** Theater tickets cost $48 per person. About how much money would tickets for 35 people cost?

Ⓐ $2,000   Ⓒ $1,200

Ⓑ $1,500   Ⓓ $200

Check My Progress

Vocabulary Check

1. The **Commutative Property of Multiplication** states that the order in which two numbers are multiplied does not change the product. Provide an example below.

   _____

2. The **Associative Property of Multiplication** states that the grouping of factors does not change the product. Provide an example below.

   _____

3. An **estimate** is an answer that is close to the exact answer. Provide an example below.

   _____

   _____

Concept Check

Multiply.

4. $\begin{array}{r} 38 \\ \times\ 30 \\ \hline \end{array}$

5. $\begin{array}{r} 52 \\ \times\ 20 \\ \hline \end{array}$

6. $\begin{array}{r} 47 \\ \times\ 10 \\ \hline \end{array}$

Estimate.

7. $\begin{array}{r} 15 \rightarrow \\ \times\ 28 \rightarrow \\ \hline \end{array}$

8. $\begin{array}{r} 71 \rightarrow \\ \times\ 51 \rightarrow \\ \hline \end{array}$

9. $\begin{array}{r} \$12 \rightarrow \\ \times\ 32 \rightarrow \\ \hline \end{array}$

10. John jogs 30 miles every week. There are 52 weeks in a year. How many miles does John jog in a year?

_____

11. Ms. Armstrong drives a total of 42 miles each day to and from work. About how many miles does Ms. Armstrong drive in 18 work days?

_____

12. A newborn alligator is 10 inches long. What is the total length of 30 newborn alligators?

_____

Brain Builders

13. The average person sends about 25 e-mails a month. About how many e-mails is this each year? Write a number model to support your answer.

_____

14. Mae estimated the product of 26 and 16 to be 400. Explain the error that Mae might have made.

_____

_____

15. Test Practice A kangaroo can travel 30 feet per jump. How far can a kangaroo travel if it jumps 14 times?

Ⓐ 420 feet   Ⓒ 52 feet

Ⓑ 320 feet   Ⓓ 42 feet

Name ....

Lesson 3
Hands On
Use the Distributive Property to Multiply

ESSENTIAL QUESTION ?
How can I multiply by a two-digit number?

You have used the Distributive Property to find a product of a two-digit number and a one-digit number.

$$3 \times 11 = 3 \times (10 + 1)$$
$$= (3 \times 10) + (3 \times 1)$$
$$= \underline{\hspace{2cm}} + \underline{\hspace{2cm}}$$
$$= \underline{\hspace{2cm}}$$

You can also use the Distributive Property to find the product of a two-digit number and a two-digit number.

Draw It 🛠 Tools

Find 12 × 15.

 Label **12** and **15** as the dimensions of the area model.

2 Separate the tens and ones of one factor. Label each part.

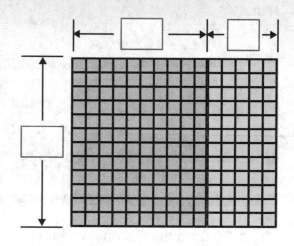

Write 15 as _____ and _____.

$12 \times 15 = 12 \times (10 + 5)$

3 Find each product. Then add.

$12 \times 15 = 12 \times (10 + 5)$

$\qquad = (12 \times 10) + (12 \times 5)$

$\qquad = \underline{\hspace{2cm}} + \underline{\hspace{2cm}}$

$\qquad = \underline{\hspace{2cm}}$

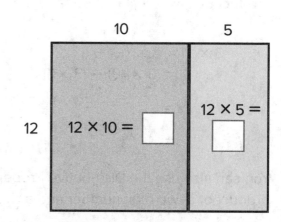

So, $12 \times 15 = \underline{\hspace{2cm}}$.

Talk About It

Processes
1. **&Practices** **7** **Identify Structure** How would you use the Distributive Property to find 12×18? Then find the product.

_____

_____

2. How would you use the Distributive Property to find 14×17? Then find the product.

_____

_____

_____

Name ....

Practice It

Draw an area model. Then use the Distributive Property to find each product.

3. Find 36 × 24.

	20	4
36	720	144

$36 \times 24 = 36 \times (20 + 4)$

$= (36 \times 20) + (36 \times 4)$

$= \underline{\hspace{1.5cm}} + \underline{\hspace{1.5cm}}$

$= \underline{\hspace{1.5cm}}$

4. Find 47 × 19.

$47 \times 19 = 47 \times (10 + 9)$

$= (47 \times \underline{\hspace{1.5cm}}) + (47 \times \underline{\hspace{1.5cm}})$

$= \underline{\hspace{1.5cm}} + \underline{\hspace{1.5cm}}$

$= \underline{\hspace{1.5cm}}$

5. Find 52 × 11.

$52 \times 11 = \underline{\hspace{1.5cm}} \times (\underline{\hspace{1.5cm}} + \underline{\hspace{1.5cm}})$

$= (\underline{\hspace{1.5cm}} \times \underline{\hspace{1.5cm}}) +$

$(\underline{\hspace{1.5cm}} \times \underline{\hspace{1.5cm}})$

$= \underline{\hspace{1.5cm}} + \underline{\hspace{1.5cm}}$

$= \underline{\hspace{1.5cm}}$

6. Find 46 × 22.

$46 \times 22 = \underline{\hspace{1.5cm}} \times (\underline{\hspace{1.5cm}} + \underline{\hspace{1.5cm}})$

$= (\underline{\hspace{1.5cm}} \times \underline{\hspace{1.5cm}}) +$

$(\underline{\hspace{1.5cm}} \times \underline{\hspace{1.5cm}})$

$= \underline{\hspace{1.5cm}} + \underline{\hspace{1.5cm}}$

$= \underline{\hspace{1.5cm}}$

Copyright © McGraw-Hill Education

Apply It

Use the Distributive Property to solve.

Processes &Practices **7.** **7** **Identify Structure** There are 15 types of animals in each part of the zoo. The zoo has 12 parts. How many types of animals are there in all?

_____

8. Write the multiplication sentence for the area model. Then find the product.

_____ × _____ = _____

Processes &Practices **9.** **3** **Find the Error** Tim drew a model to find 11×25. Find and correct his mistake.

$200 + 25 = 225$

_____

_____

10. Why is the Distributive Property appropriate for two-digit multiplication? Explain.

_____

_____

_____

Name ....

Homework Helper

Need help? ⟋ **connectED.mcgraw-hill.com**

Find 26 × 25.

An area model can be used to represent the factors.
The tens and ones of one factor are separated.

Find each product. Then add.

$$26 \times 25 = 26 \times (20 + 5)$$
$$= (26 \times 20) + (26 \times 5)$$
$$= 520 + 130$$
$$= 650$$

So, 26 × 25 = 650.

Practice

Draw an area model. Then use the Distributive Property to find each product.

1. 73 × 34 = _____

$$73 \times 34 = 73 \times (30 + 4)$$
$$= (73 \times \text{\_\_\_\_\_}) + (73 \times \text{\_\_\_\_\_})$$
$$= \text{\_\_\_\_\_} + \text{\_\_\_\_\_}$$
$$= \text{\_\_\_\_\_}$$

2. 82 × 22 = _____

$$82 \times 22 = 82 \times (20 + 2)$$
$$= (82 \times \text{\_\_\_\_\_}) + (82 \times \text{\_\_\_\_\_})$$
$$= \text{\_\_\_\_\_} + \text{\_\_\_\_\_}$$
$$= \text{\_\_\_\_\_}$$

Draw an area model. Then use the Distributive Property to find each product.

3. $18 \times 39 =$ _____

$18 \times 39 =$ _____ $\times$ (_____ + _____)

$=$ (_____ $\times$ _____) +

(_____ $\times$ _____)

$=$ _____ + _____

$=$ _____

Problem Solving

4. There are 48 nails in one box. How many nails are in 17 boxes?

_____ nails

$17 \times 48 =$ _____ $\times$ (_____ + _____)

$=$ (_____ $\times$ _____) +

(_____ $\times$ _____)

$=$ _____ + _____

$=$ _____

5. Each notebook has 64 pages. How many total pages are there in 33 notebooks?

_____ pages

$33 \times 64 =$ _____ $\times$ (_____ + _____)

$=$ (_____ $\times$ _____) +

(_____ $\times$ _____)

$=$ _____ + _____

$=$ _____

6. Each jar contains 55 buttons. There are 16 jars on the shelf. How many buttons are there all together?

_____ buttons

Name _____

ESSENTIAL QUESTION
How can I multiply by a two-digit number?

 Math in My World Watch Tutor

Example 1

A coyote travels 27 miles each hour. How far can a coyote travel in 12 hours?

Find 27 × 12.

One Way Use partial products.

Draw an area model. Separate each factor into tens and ones. Multiply. Then add the partial products.

200 + 40 + 70 + 14 = _____

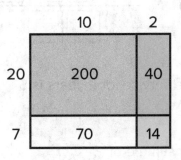

	10	2
20	200	40
7	70	14

Another Way Use paper and pencil.

1 **Multiply the ones.**

$$\begin{array}{r} 27 \\ \times\ 2 \\ \hline 54 \end{array}$$

7 × 2 = 14
Regroup the tens.
2 tens × 2 ones = 4 tens
4 tens + 1 ten = 5 tens

2 **Multiply the tens.**
27 × 1 ten = 27 tens, or 270

3 **Add the products.**
54 + 270 = 324

$$\begin{array}{r} 2\quad 7 \\ \times 1\quad 2 \\ \hline \end{array}$$

+

27 × 12 = _____

So, a coyote can travel _____ miles in 12 hours.

Online Content at **connectED.mcgraw-hill.com**

Example 2

Heidi's monthly bills are shown. How much does she spend on her cell phone service in 2 years?

Write an equation to help you solve the problem.

$$\$38 \times 12 \times 2 = m$$

Heidi's cell phone bill each month

12 months in one year

2 years

Monthly Bills	
Cable	$55
Cell phone	$38
Movie club	$21
Water	$93

You know that $12 \times 2 = 24$. So, you need to find $\$38 \times 24$.

Estimate _____ × _____ = _____

$$\begin{array}{r} \$3\ \ 8 \\ \times 2\ \ 4 \\ \hline \end{array}$$

1 Multiply the ones. ⟶ $ ☐☐☐

2 Multiply the tens. ⟶ + $ ☐☐☐

3 Add the products. ⟶ $ ☐☐☐

So, the cost of cell phone service for 2 years is $ _____ .

Check

_____ is close to the estimate of _____ .

Explain the steps needed to find the product of 56 and 23.

Guided Practice

Multiply.

1.

$$\begin{array}{r} 3\ \ 5 \\ \times 2\ \ 4 \\ \hline \end{array}$$

☐☐☐ Multiply the ones.

+ ☐☐☐ Multiply the tens.

☐☐☐ Add.

The area model shows that

$$600 + 120 + 100 + 20 = \underline{\hspace{2cm}} .$$

Area model: 30 and 5 across top; 20 and 4 down side; cells show 600, 100, 120, 20.

Independent Practice

Multiply. Use the area model to check.

2.
19
× 15

3.
42
× 38

4. $54
× 51

5. $74
× 63

Multiply.

6.
47
× 24

7.
64
× 46

8. 83 × 67 = _____

9. 91 × 78 = _____

Problem Solving

10. A greyhound dog can jump a distance of 27 feet. How many feet will a greyhound travel if it jumps 12 times?

_____

11. The fourth-grade students at Tremont School received a ribbon if they read 50 books during the school year. The school gave out 69 ribbons at the end of the year. How many books did the students read in all?

_____

12. **Processes &Practices** **2** **Use Symbols** Each day, enough paper is recycled in the United States to fill 15 miles of train boxcars. How many miles of boxcars could be filled in 5 weeks? Complete the equation to help you solve the problem.

15 × 5 × _____ = *b*

_____

Brain Builders

13. **Processes &Practices** **3** **Which One Doesn't Belong?** Circle the multiplication problem that does not belong with the other three. Explain and rewrite the problem so that it does belong.

22	$45	37	$66
×15	×28	×18	×25
___	___	___	___

_____

14. **Building on the Essential Question** Why can't the product of two 2-digit numbers ever be two digits? Explain.

_____

MY Homework

Homework Helper

Need help? ↗ connectED.mcgraw-hill.com

Find 29 × 56.

1 Multiply the ones.

2 Multiply the tens.

3 Add the products.

$$
\begin{array}{r}
56 \\
\times\ 29 \\
\hline
504 \\
+\ 1{,}120 \\
\hline
1{,}624
\end{array}
$$

504 ← 9 × 56

+ 1,120 ← 20 × 56

	50	6
20	1,000	120
9	450	54

So, 29 × 56 = 1,624.

Practice

Multiply.

1.
$$
\begin{array}{r}
26 \\
\times\ 35 \\
\hline
\end{array}
$$

2.
$$
\begin{array}{r}
\$46 \\
\times\ 35 \\
\hline
\end{array}
$$

3.
$$
\begin{array}{r}
79 \\
\times\ 73 \\
\hline
\end{array}
$$

4.
$$
\begin{array}{r}
73 \\
\times\ 51 \\
\hline
\end{array}
$$

5.
$$
\begin{array}{r}
59 \\
\times\ 47 \\
\hline
\end{array}
$$

6.
$$
\begin{array}{r}
94 \\
\times\ 61 \\
\hline
\end{array}
$$

Multiply.

7.   44
       × 87

8.   77
       × 22

Problem Solving

9. Mrs. Taylor gave each of her students 75 pieces of paper at the beginning of the school year. If there are 32 students in her class, how many pieces of paper did she give out all together?

10. Mr. Matthews gives each of his 32 students 15 minutes to present a book report to the class. How many minutes will it take for all of his students to present their reports?

Brain Builders

11. **Processes &Practices** **2** **Use Symbols** George collects 25 baseball cards each month. Annie collects 12 baseball cards each month. How many cards will they have at the end of one year? Complete the equations to help you solve the problem.

$25 \times \underline{\hspace{1.5cm}} = n$

$12 \times \underline{\hspace{1.5cm}} = m$

12. The office building is 48 floors high. Half of the floors have 18 windows each and half of the floors have 36 windows each. How many windows does the building have in all?

13. **Test Practice** There are 26 rows of bleachers in the school gym. At the pep rally, there were 17 students sitting in each of the first 24 rows and 10 students sitting in each of the remaining rows. How many students were there in all?

    Ⓐ 27 students   Ⓒ 428 students

    Ⓑ 408 students   Ⓓ 442 students

Name ....

ESSENTIAL QUESTION
How can I multiply by a two-digit number?

Sometimes, problems require more than one operation to solve. An **operation** is a mathematical process such as addition, subtraction, multiplication, or division.

Math in My World

Example 1

Francis earns $8 each week walking dogs. She spends $3 each week and saves the rest. There are 52 weeks in a year. How much will she have saved at the end of the year?

You need to find ($8 − $3) × 52. The operations that are needed for this problem are subtraction and multiplication.

Estimate ($8 − $3) × 52   Round 52 to 50.

$5 × 50 = $250

Subtract. ($8 − $3) × 52

$5 × 52

Multiply.
```
    5   2
×       5
 ☐  ☐  ☐
```

So, Francis will have saved $ _____ .

Check

$ _____ is close to the estimate $250. So, the answer is correct.

You can use a variable to represent unknown quantities.

Example 2

Coach Murphy bought three boxes of trophies. Each box has 45 soccer trophies. He also bought 15 tennis trophies and some golf trophies. There are a total of 170 trophies. Write an equation to describe the number of trophies that Coach Murphy bought. How many golf trophies did Coach Murphy buy?

soccer trophies   tennis trophies   golf trophies

$(3 \times 45) + 15 + ? = 170$ trophies

Replace the unknown quantity with a variable.

$(3 \times 45) + 15 + g = 170$

Write an equation.

Solve for the unknown quantity.

1 Multiply.

$(3 \times 45) + 15 + g = 170$

$\boxed{} + 15 + g = 170$

2 Add.

$135 + 15 + g = 170$

$\boxed{} + g = 170$

3 Subtract.

$150 + g = 170$

Subtract 150 from 170 to find the value of g.

$170 - 150 = \underline{}$

170	
150	g

So, Coach Murphy bought _____ golf trophies.

Guided Practice

1. Karrie has 20 bags of prizes. Each bag has 4 prizes. She also has a red bag with 13 prizes and a blue bag with the rest of the prizes. She has a total of 100 prizes. How many prizes are in the blue bag? Write an equation. Use a variable for the unknown number.

Talk MATH

Why are variables used?

_____

_____

Independent Practice

Algebra Write an equation for each problem. Solve.

2. Each small dog at doggy day care weighs 35 pounds. Each large dog weighs 60 pounds. There are 4 small dogs and 6 large dogs. How much do the dogs weigh all together?

3. Suzie has track practice for 1 hour on Tuesday and 2 hours on Thursday. How many hours does Suzie go to track practice in 15 weeks?

Algebra Write an equation for each problem. Use a variable for the unknown number. Solve.

4. Warren, Lisa, and Tina went to the carnival. The table shows the number of points Warren won at each carnival game.

Game	Points
Banana Bowling	24
Toss & Trade	16
Rabbit Race	10

Lisa won the same number of points as Warren. Warren, Lisa, and Tina won a total of 225 points. How many points did Tina win?

5. Mark bought four hats that each cost $8. He also bought a shirt that cost $14 and a pair of jeans. He spent a total of $68. How much did the jeans cost?

Problem Solving

Use a number cube to complete each number puzzle.

6. Processes &Practices **Keep Trying** Roll a number cube four times.

The numbers rolled are: _____ ,

_____ , _____ , and _____ .

Write the numbers in the boxes below. Use each number once.
Try to create the greatest number possible.

(▢ × ▢) + ▢ − ▢ = _____

7. Roll a number cube four times.

The numbers rolled are: _____ , _____ , _____ ,

and _____ .

Write the numbers in the boxes below. Use each number once.
Try to create the greatest number possible.

▢ + (▢ × ▢) − ▢ = _____

Brain Builders

8. Processes &Practices **Make Sense of Problems** A bus has 15 rows of seats.
Each row has 4 seats. At the first stop, 25 people get on the bus. At the
second stop, 3 people get off of the bus, and 12 people get on the bus.
How many empty seats are there after the second stop?

_____

_____

9. **Building on the Essential Question** How can I use equations to
model real-world problems? How can I use equations to check my
answers?

_____

Name ....

MY Homework

Homework Helper

Need help? 🔍 connectED.mcgraw-hill.com

**A store had 3 baskets. Each basket had 62 necklaces. In the
morning, 25 necklaces were sold. In the afternoon, some necklaces
were returned. At the end of the day, there were 166 necklaces.
How many necklaces were returned?**

Write an equation to represent the problem.

baskets of necklaces	−	necklaces sold	+	necklaces returned	= total
↓		↓		↓	↓
(3×62)	−	25	+	n	= 166

↑
Use a variable.

Solve for the unknown quantity.

1. Multiply.

$(3 \times 62) - 25 + n = 166$

$\boxed{} - 25 + n = 166$

2. Subtract.

$186 - 25 + n = 166$

$\boxed{} + n = 166$

3. Use mental math. $161 + n = 166$

$161 + 5 = 166$

So, 5 necklaces were returned.

⊢------- 166 -------⊣	
161	n

Practice

1. Gina works at a diner. She earns $6 each hour plus tips. In one
   week, she worked 37 hours a week and earned $43 in tips. How
   much did she make all together? Write an equation. Use a variable
   for the unknown. Solve.

_____

Copyright © McGraw-Hill Education

Problem Solving

Write an equation for each problem. Use a variable for the unknown number. Solve.

2. **Processes &Practices** **Use Number Sense** It costs $45 to rent a car each day. There is also a fee of $12. How much does it cost to rent a car for 5 days, including the fee?

3. A climbing gym charges $10 to climb each day. A pair of climbing shoes costs $84. It costs $169 to buy 6 days of climbing, one pair of climbing shoes, and one harness. How much does a harness cost?

Brain Builders

4. A travel agency charges $64 for each bus ticket and $82 for each train ticket. Aaron has $520 to purchase tickets. If he buys 3 bus tickets, how many train tickets can Aaron buy?

Vocabulary Check

5. Label each part of the equation. Write *operation* or *variable*.

$$n - 100 + r = 54$$

6. **Test Practice** There are three shelves. Each shelf has 28 books. There is also a stack of some more books. There are a total of 85 books. Which equation represents this situation?

Ⓐ $(3 \times 28) + b = 85$ Ⓒ $(3 \times 28) + 85 = b$

Ⓑ $(3 + 28) \times b = 85$ Ⓓ $(3 + 28) \times 85 = b$

Lesson 6
Problem-Solving Investigation
STRATEGY: Make a Table

Learn the Strategy

 Watch Tutor

Each roller coaster car holds 18 people. Every minute, a new car is filled. Make a table to find how many people can ride in 60 minutes.

 Understand

What facts do you know?

There are _____ people per car.

What do you need to find?

the number of _____ who can ride in _____ minutes

 Plan

I can make a table to find the number of people who can

ride in _____ minutes.

Solve

Start by finding the product of 18 and 10. 18 × 10 = 180

Minutes	10	20	30	40	50	60
Passengers	180					

So _____ people can ride the roller coaster in 60 minutes.

 Check

Does your answer make sense? Explain.

Yes. _____ × _____ = _____

There are 20 sea lions in a circus. Each sea lion can juggle 5 balls at a time. How many balls will the sea lions need for their act if they all perform at the same time?

 Understand

What facts do you know?

_____

_____

What do you need to find?

_____

_____

 Plan

_____

_____

 Solve

_____

_____

4 Check

Does your answer make sense? Explain.

_____

Apply the Strategy

Solve each problem by making a table.

1. A page from Dana's album is shown. Dana puts the same number of stickers on each page. She has 30 pages of stickers. How many stickers does she have in all?

   Dana has _____ stickers in all.

2. West Glenn School has 23 students in each class. There are 6 fourth grade classes. About how many fourth grade students are there in all?

   There are about _____ students. I solved this by

   _____

   _____

   _____

Brain Builders

Processes &Practices 3 **Draw a Conclusion** Evita had homework four
3. nights this week. She spent 30 minutes on math and 20 minutes on her English essay each night. Write a real-world problem using this information. Then solve.

   _____

   _____

   _____

4. Ling exercises for 45 minutes 2 times a day. If she keeps up this schedule for 30 days, how many hours will she exercise in all?

   _____

Review the Strategies

Use any strategy to solve each problem.

- Make a table.
- Find an estimate or exact answer.
- Find reasonable answers.
- Draw a diagram.

5. Corey and his 2 friends earn $12 each for doing yard work. How much money will they get paid all together if they work on 5 yards? Make a table.

3 friends × $12 a yard = $36

1 yard	2 yards			
$36				

6. A lemur sleeps 16 hours each day. A sloth sleeps 4 more hours each day than a lemur. How many total hours of sleep do a lemur and a sloth get during two days?

7. A lizard eats 6 crickets each day. How many crickets does it eat in 13 weeks?

Processes &Practices **5** **Use Math Tools** Write a real-world problem that would involve making a table to find the answer.

8.

9. Pete spends 30 minutes a night reading. How many hours does he spend reading in a 30-day month?

Name

MY Homework

Homework Helper

Need help? ⟋ connectED.mcgraw-hill.com

**The cafeteria serves breakfast to 48 students each morning.
In a week of 5 school days, how many times is breakfast served?**

1 Understand

I know that 48 students are served breakfast each morning for 5 days.

2 Plan

I can make a table to find 48 × 5.

3 Solve

Day	1	2	3	4	5
Breakfast served	48	96	144	192	240

So, breakfast is served 240 times in one week.

4 Check

Multiply 5 × 48.
5 × 48 = 240

 ## Problem Solving

1. Mrs. Shelley's class is reading *The Lion, the Witch, and the Wardrobe*. If they read 16 pages every week, how many pages can they read in 5 weeks? Solve the problem by making a table.

Week				
Pages				

Solve each problem by making a table.

**Processes
&Practices** ⑧ **Look for a Pattern** Fiona finds 1 shell on her first day at the beach. Every day that week, she finds twice as many shells as she did the day before. How many shells does Fiona find on the seventh day?

_____

3. The parking garage holds 300 cars on each level. There are 4 levels in the garage. How many cars can the parking garage hold in all?

_____

Brain Builders

4. Jonah sets the table for breakfast and dinner every Monday, Wednesday, and Friday. How many times does he set the table in six weeks?

_____

5. Tony and his brother have 20 thank-you cards to write. If they each write 2 per day, how many days will it take them to finish?

_____

6. At a dance recital, each dancer performs for 13 minutes. There are 6 dancers to perform. The reception that follows the dancers is 20 minutes long. How long is the event?

Dancers						
Minutes						

_____

Vocabulary Check

Using the word bank below, write the correct word in each blank.

Associative Property   **Commutative Property**   **Distributive Property**

operation   **partial products**

1. This property states that the order in which two numbers are multiplied does not change the product.

$$23 \times 11 = 11 \times 23$$

_____

2. In the equation $32 \times 10 = 320$, the multiplication symbol represents this process.

_____

3.
$$\begin{array}{r} 18 \\ \times\ 27 \\ \hline 126 \\ +360 \\ \hline 486 \end{array}$$

[   ]

4. This property states that the grouping of the factors does not change the product.

_____

5. This property states that multiplying a sum by a number is the same as multiplying each addend by the number and then adding the products.

$$2 \times 12 = 2 \times (10 + 2)$$
$$= (2 \times 10) + (2 \times 2)$$
$$= 20 + 4$$
$$= 24$$

_____

Concept Check

Multiply.

6.   90
   × 90

7.   34
   × 80

8. $28
   × 40

9. $45
   × 30

**Estimate. Circle whether the estimate is _greater than_ or
less than the actual product.**

10. $24 ⟶
     × 31 ⟶ × _____

greater than

less than

11.   48 ⟶
     × 89 ⟶ × _____

greater than

less than

12.   37 ⟶
     × 66 ⟶ × _____

greater than

less than

13. $52 ⟶
     × 84 ⟶ × _____

greater than

less than

Multiply.

14.   63
    × 46

15.   26
    × 34

16. $72
    × 49

17. $55
    × 41

Problem Solving

18. A football coach orders 30 jerseys for his football team. The jerseys cost $29 each. What is the total cost of the jerseys?

19. Julio scores 18 points in each basketball game. If there are 14 games in a season and Julio continues to score 18 points each game, how many points will Julio score?

20. There are 30 students in each class. There are 27 classrooms. How many students are there all together?

Brain Builders

21. Tamara makes $12 an hour. She worked 3 hours each weekday and 13 hours over the weekend. About how much money will she make?

22. Austin's allowance is $15 per week. He spends $4 each week on baseball cards. How much money will Austin have at the end of 12 weeks?

23. **Test Practice** A jumbo can of vegetables has 36 servings. How many servings of vegetables are in 18 cans?

  Ⓐ 648 servings   Ⓒ 608 servings

  Ⓑ 324 servings   Ⓓ 54 servings

Use what you learned about multiplying with
two-digit numbers to complete the graphic organizer.

Write the
Example

Real-World
Problem

**ESSENTIAL
QUESTION**

How can I
multiply by a two-digit
number?

Vocabulary

Estimate

Reflect on the ESSENTIAL QUESTION Write your answer below.

Performance Task

Brain Builders

Pet Rescue

Kate and Ju help at the animal rescue. The rescue was started after an emergency, and now is the home to all kinds of animals.

Show all your work to receive full credit.

Part A

Kate and Ju asked a dog food company for donations. The company offered to donate 75 pounds of dog food each month for two years. Complete the area model to find the total amount of dog food the company will donate. Then use the model to find the total.

Part B

The company also agreed to donate 82 pounds of cat food to the rescue each month for two years. Explain how you can use your answer from Part A to find the total number of pounds of cat food that will be donated. Then state the total.

Part C

The rescue has 48 cats in foster homes. The director asked Kate to pack a box of canned cat food for each foster home. Each box will contain 34 cans. Kate does the multiplication shown to find out how many cans she will need in all. Use place value to explain the error that Kate made. Then correct the error and find the total number of cans she will need.

$$
\begin{array}{r}
48 \\
\times\ 34 \\
\hline
192 \\
+144 \\
\hline
336
\end{array}
$$

Part D

Ju is helping build an outdoor dog run for the 28 dogs that live at the rescue. Each dog should have at least 56 square feet of space. Write an equation that could be used to find s, an estimate of the total number of square feet the entire dog pen should enclose. Then solve the equation.

Part E

Is your estimate from Part D greater than or less than the actual number of square feet the dogs need? Explain. In this case, explain whether it would be better for the estimate to be greater than or less than the actual number.

Chapter

6 Divide by a One-Digit Number

ESSENTIAL QUESTION

How does division affect numbers?

Let's Travel!

Watch a video!

MY Chapter Project

Division Game Day

Your teacher will put you into a group and assign you to a game. Follow the instructions of the game you were assigned below. After some time, your teacher will tell your group to switch games.

Musical Quotients

1. Set up six hula hoops. When the music plays, walk in a circle around the hula hoops. When the music stops, quickly divide yourselves evenly among the hula hoops. There should be an equal number of your classmates in each hoop. Any "remainder" students are out of the game.

2. After each round, write the division problem on the board with its answer. For example, if there are 16 students in your group and 6 hula hoops, the division problem is $16 \div 6 = 2$ R4.

3. Remove one hula hoop and then begin the game again.

Division Memory

1. Take two index cards. Write a division problem on one card, and its answer on the other card. Make two piles of cards, one for the problems and one for the answers.

2. Shuffle both sets of cards and place each set face down on a table.

3. Divide yourselves into three teams. Each team will take turns flipping one card from the problem set and one card from the answer set. If the problem matches the answer, the team gets to keep the cards. If the cards do not match, place them back on the table face down.

4. The game is finished when all of the cards are gone. The team with the most cards wins the game.

Am I Ready?

Subtract.

1.
$$\begin{array}{r} 1025 \\ -6 \\ \hline \end{array}$$

2.
$$\begin{array}{r} 2{,}642 \\ -8 \\ \hline \end{array}$$

3.
$$\begin{array}{r} 3{,}467 \\ -29 \\ \hline \end{array}$$

4. $7{,}024 - 15 =$ _____

5. $1{,}331 - 17 =$ _____

6. $6{,}050 - 23 =$ _____

7. There are 1,080 pages in Gerardo's book. He has read 1,038 pages. How many pages are left to read?

_____

Divide.

8. $2\overline{)16}$

9. $3\overline{)9}$

10. $3\overline{)24}$

11. $35 \div 5 =$ _____

12. $48 \div 8 =$ _____

13. $56 \div 7 =$ _____

14. Sharon has $32. She wants to buy CDs that cost $8 each. How many can she buy?

_____

Shade the boxes to show the problems you answered correctly.

How Did I Do? ➤

1	2	3	4	5	6	7	8	9	10	11	12	13	14

Online Content at ↗ connectED.mcgraw-hill.com

MY Math Words

Vocab
abc

Review Vocabulary

dividend   divisor   quotient

Making Connections

Read the word problem. Use the review vocabulary to describe what each number represents.

> A local pond is home to 36 Canada geese.
> They live in groups. Each group has 9 geese.
> How many Canada geese groups are there?

The total number of Canada geese is _____. This number represents the _____.	There are _____ Canada geese in each group. This number represents the _____.	The number of groups living in the pond represents the _____.

Write and solve a division sentence about the word problem.
Circle the quotient.

_____

Lesson 6–2

compatible numbers

Estimate 4,588 ÷ 9.

4,500 ÷ 9

compatible numbers

Lesson 6–8

partial quotients

4)624
−500   125 ← partial quotient
124
−100   25 ← partial quotient
24
−24   6 ← partial quotient
125 + 25 + 6 = 156

Lesson 6–3

remainder

74 ÷ 4 = 18 R2

Ideas for Use

- Work with a partner to name the part of speech of each word. Consult a dictionary to check your answers.

- Use the blank cards to write your own vocabulary cards.

A dividing method in which the dividend is separated into sections that are easy to divide.

How can the meaning of *partial* help you remember this vocabulary word?

_____

_____

Numbers in a problem that are easy to work with mentally.

How can you use basic facts to estimate a quotient?

_____

_____

The number that is left after one whole number is divided by another.

Use the model to write a division equation.

MY Foldable

FOLDABLES Follow the steps on the back to make your Foldable.

✂

Divide

2)42 3)37 3)38 R □ □ □

Multiply

Subtract Compare

Bring Down Start Over

Remainder?

Copyright © McGraw-Hill Education

FOLDABLES®
Study Organizer

1

2

3

B its

S maller

M akes

D ividing

Name ....

Lesson 1
Divide Multiples of 10, 100, and 1,000

You use place value and patterns to divide dividends that are multiples of 10, 100, and 1,000.

 Math in My World

Example 1

Anita's family went on vacation to an amusement park. The park has 5 entrances. 1,500 people entered the park and separated into equal lines. How many people are in each line?

Divide 1,500 people into 5 equal groups.

One Way   Use a multiplication pattern.

$5 \times 3 = 15$ $\longrightarrow$ $15 \div 5 = 3$

$5 \times 30 = 150$ $\longrightarrow$ $150 \div 5 = 30$

$5 \times 300 = 1,500$ $\longrightarrow$ $1,500 \div 5 =$ _____

Another Way   Use a basic fact and place value.

$15 \div 5 = 3$ ⟵ basic fact

$150 \div 5 = 30$ ⟵ 150 is 10 times as big as 15. So, the quotient 30, is 10 times as big as 3.

$1,500 \div 5 =$ _____ ⟵ 1,500 is 100 times as big as 15. So, the quotient is 100 times as big as 3.

So, there are _____ people in each line.

Copyright © McGraw-Hill Education

Example 2

Find the quotient of 2,400 and 4.

Find $2,400 \div 4$.

One Way   Use a multiplication pattern.

$4 \times 6 = 24$ $\longrightarrow$ $24 \div 4 = 6$

$4 \times 60 = 240$ $\longrightarrow$ $240 \div 4 = 60$

$4 \times 600 = 2,400$ $\longrightarrow$ $2,400 \div 4 = \underline{\hspace{1cm}}$

Another Way   Use a basic fact and place value.

$24 \div 4 = 6$ $\longleftarrow$ [basic fact]

$240 \div 4 = 60$ $\longleftarrow$ [$240 = 10 \times 24$. So, $60 = 10 \times 6$.]

$2,400 \div 4 = \underline{\hspace{1cm}}$ $\longleftarrow$ [$2,400 = 100 \times 24$. So, the quotient is 100 times as big as 6.]

So, $2,400 \div 4 = \underline{\hspace{1cm}}$.

Check

Use multiplication to check division.

$2,400 \div 4 = \underline{\hspace{1cm}}$

$\underline{\hspace{1cm}} \times 4 = 2,400$

What basic fact will help you find the quotient of 4,200 and 7?

Guided Practice

Complete each set of patterns.

1. $12 \div 4 = \underline{\hspace{1cm}}$ **2.** $36 \div 9 = \underline{\hspace{1cm}}$

   $120 \div 4 = \underline{\hspace{1cm}}$ $360 \div 9 = \underline{\hspace{1cm}}$

   $1,200 \div 4 = \underline{\hspace{1cm}}$ $3,600 \div 9 = \underline{\hspace{1cm}}$

Divide. Use patterns and place value.

3. $\$400 \div 2 = \underline{\hspace{1cm}}$

4. $1,600 \div 4 = \underline{\hspace{1cm}}$

Name ....

Independent Practice

Complete each set of patterns.

5. $12 \div 2 =$ _____

$120 \div 2 =$ _____

$1,200 \div 2 =$ _____

6. $54 \div 9 =$ _____

$540 \div 9 =$ _____

$5,400 \div 9 =$ _____

7. $\$36 \div 4 =$ _____

$\$360 \div 4 =$ _____

$\$3,600 \div 4 =$ _____

8. $42 \div 6 =$ _____

$420 \div 6 =$ _____

$4,200 \div 6 =$ _____

9. $\$28 \div 7 =$ _____

$\$280 \div 7 =$ _____

$\$2,800 \div 7 =$ _____

10. $\$72 \div 8 =$ _____

$\$720 \div 8 =$ _____

$\$7,200 \div 8 =$ _____

Divide. Use patterns and place value.

11. $200 \div 5 =$ _____

12. $\$600 \div 3 =$ _____

13. $900 \div 3 =$ _____

14. $800 \div 2 =$ _____

15. $\$1,400 \div 7 =$ _____

16. $4,500 \div 5 =$ _____

17. $\$3,500 \div 5 =$ _____

18. $6,300 \div 9 =$ _____

19. $\$6,400 \div 8 =$ _____

20. $1,600 \div 8 =$ _____

21. $5,400 \div 6 =$ _____

22. $\$8,100 \div 9 =$ _____

Copyright © McGraw-Hill Education

Lesson 1 Divide Multiples of 10, 100, and 1,000   **331**

Problem Solving

Animals migrate due to factors such as climate and food availability. The table shows a few migration distances.

Migration	
Animals	Distance (in miles)
Caribou	2,400
Desert locust	2,800
Green sea turtle	1,400

23. Suppose a group of green sea turtles travels 7 miles a day. How many days will the migration take?

24. **Processes &Practices** **4** **Model Math** A herd of caribou migrated the distance shown in 8 months. If they traveled the same distance each month, how many miles did the herd travel each month?

Brain Builders

25. **Processes &Practices** **5** **Use Mental Math** Using mental math, order the quotients, 1,500 ÷ 3, 2,400 ÷ 8, and 4,800 ÷ 12, from least to greatest. Explain your reasoning.

26. **Processes &Practices** **1** **Plan Your Solution** Complete the equation. Explain your reasoning.

$$\boxed{},80\boxed{} \div 6 = \boxed{}\boxed{}\boxed{}$$

27. **Building on the Essential Question** Why are basic facts needed when dividing large numbers?

Name ....

MY Homework

Homework Helper

Need help? 🔗 connectED.mcgraw-hill.com

Find 2,700 ÷ 9.

The dividend, 2,700, is a multiple of 100. You can use a basic fact and place value to solve.

$27 \div 9 = 3$ ◄── This is the basic fact.

$270 \div 9 = 30$ ◄── See the pattern: 270 is 10 × 27, and 30 is 10 × 3.

$2,700 \div 9 = 300$ ◄── Continue the pattern: 2,700 is 100 × 27, and 300 is 100 × 3.

So, $2,700 \div 9 = 300$.

Practice

Complete each set of patterns.

1. $24 \div 3 =$ _____

$240 \div 3 =$ _____

$2,400 \div 3 =$ _____

2. $32 \div 8 =$ _____

$320 \div 8 =$ _____

$3,200 \div 8 =$ _____

3. $45 \div 5 =$ _____

$450 \div 5 =$ _____

$4,500 \div 5 =$ _____

4. $56 \div 8 =$ _____

$560 \div 8 =$ _____

$5,600 \div 8 =$ _____

Divide. Use patterns and place value.

5. $1,000 \div 2 =$ _____

6. $500 \div 10 =$ _____

7. $300 \div 5 =$ _____

8. $2,100 \div 3 =$ _____

9. $7,200 \div 9 =$ _____

10. $\$2,000 \div 4 =$ _____

11. $4,200 \div 7 =$ _____

12. $\$2,400 \div 6 =$ _____

Problem Solving

Copyright © McGraw-Hill Education

13. **Processes &Practices** 5 **Use Math Tools** The electronics store sold 4 laptop computers in one day. The total cost of the laptops was $3,600. If each laptop cost the same amount, how much did each laptop cost?

_____

14. The Thompson family is traveling 1,500 miles to visit relatives. They plan to drive the same distance each day. If the Thompsons make the trip in 3 days, how far will they drive each day?

_____

If they make the trip in 5 days, how far will they drive each day?

Brain Builders

15. Linus has 58 baseball cards and 102 football cards he wants to give to his 4 cousins. If he divides the cards equally, how many cards will he give to each cousin?

_____

16. Last year, Carlotta earned $1,200 babysitting for a family with 2 children. Carlotta charges $3 per hour per child. What is the total number of hours Carlotta spent babysitting for the family last year?

_____

17. **Test Practice** On a trip to New York City, 8 people spent a total of $900 on transportation and $1,500 on hotel rooms. If they shared the cost equally, how much did each person spend?

Ⓐ $400   Ⓒ $40

Ⓑ $300   Ⓓ $30

Name ....

Lesson 2
Estimate Quotients

ESSENTIAL QUESTION ?
How does division affect numbers?

There are different ways to estimate quotients. One way is to use compatible numbers. **Compatible numbers** are numbers that are easy to compute mentally.

Math in My World

Example 1

Circuses have been around for more than 200 years. They sometimes travel by train. Suppose a circus travels 642 miles in 8 hours. Estimate the quotient of 642 and 8 to find about how many miles per hour the train travels.

Estimate $642 \div 8$.

$$642 \div 8$$

642 is close to 640.

640 and 8 are compatible numbers because they are easy to divide mentally.

$$640 \div 8 = \underline{\hspace{2cm}}$$

> **Helpful Hint**
> 64 and 8 are members of a fact family.
> $$8 \times 8 = 64$$
> $$64 \div 8 = 8$$

So, the train is traveling about _____ miles per hour.

Copyright © The McGraw-Hill Companies, Inc.   Ken Cavanagh/McGraw-Hill Education

Online Content at connectED.mcgraw-hill.com

Lesson 2 335

Example 2

Isabella has 6 tea sets in her collection. The collection is worth $1,168. Each tea set is worth the same amount of money. About how much is each tea set worth?

You need to estimate $1,168 ÷ 6.

One Way Use compatible numbers.

$1,168 ÷ 6

$1,168 is close to $1,200. $1,200 and 6 are compatible numbers because they are easy to divide mentally.

$1,200 ÷ 6 = $ _____

Helpful Hint

12 ÷ 6 = 2, so
1,200 ÷ 6 = 200

Another Way Use a basic fact and place value.

$1,168 ÷ 6 ← What basic multiplication fact is close to the numbers in the problem?

6 × 2 = 12

6 × 20 = 120

6 × _____ = 1,200

So, each tea set is worth about _____ .

Check

Use multiplication to check division.

1,200 ÷ 6 = _____

_____ × 6 = 1,200

Explain how to estimate
$4,782 ÷ 6.

Guided Practice

1. Estimate. Check your estimate using multiplicaton.

161 ÷ 4

_____ ÷ _____ = _____

Check: _____ × _____ = _____

Independent Practice

Estimate. Check your estimates using multiplication.

2. $123 \div 3$

3. $\$244 \div 6$

4. $162 \div 2$

5. $345 \div 7$

6. $538 \div 6$

7. $415 \div 6$

8. $\$1,406 \div 7$

9. $2,431 \div 8$

10. $\$2,719 \div 9$

Algebra Use mental math to find an estimate of the unknown number.

11. $4,187 \div 7 = f$

12. $\$7,160 \div c = \$800$

13. $8,052 \div 9 = t$

f is about _____

c is about _____

t is about _____

Problem Solving

Use the following information for Exercises 14 and 15. Hut hiking involves hiking and spending the night in huts.

14. The total cost for 5 members in the Valdez family to hut hike for 6 days is $2,475. About how much does it cost for each family member?

_____

15. Ricardo needs to climb a 361-foot hill to get to the next hut. About how many yards away is he from the next hut? (*Hint:* 3 feet = 1 yard)

_____

16. **Processes &Practices** **4** **Model Math** Terrence earned 806 points on 9 tests. If he earned about the same number of points on each test, about how many points did he earn on each test?

_____

Brain Builders

17. A farm has 8 rows of beans. There are a total of 1,800 bean plants. Two rows each have 300 bean plants, and the remaining rows have a different, equal number of plants. How many bean plants are in the other six rows?

_____

_____

18. **Processes &Practices** **1** **Make a Plan** The estimated quotient of a division sentence is 200. Explain how to find a possible division sentence.

_____

_____

19. **Building on the Essential Question** How can you estimate quotients?

_____

_____

Name

Homework Helper

Need help? connectED.mcgraw-hill.com

Estimate 122 ÷ 3.

Find compatible numbers, or numbers that are easy to divide mentally.

122 is close to 120. 120 and 3 are compatible numbers because they are easy to divide mentally.

Divide using the compatible numbers. 120 ÷ 3 = 40

Check

Use multiplication. 3 × 40 = 120

So, a good estimate for 122 ÷ 3 is 40.

Helpful Hint
12 ÷ 3 = 4, so
120 ÷ 3 = 40

Practice

Estimate. Check your estimates using multiplication.

1. 184 ÷ 9	**2.** $149 ÷ 5	**3.** 241 ÷ 8
4. $422 ÷ 6	**5.** 637 ÷ 8	**6.** $3,611 ÷ 6
7. 1,175 ÷ 4	**8.** 5,421 ÷ 9	**9.** $2,782 ÷ 7

Copyright © McGraw-Hill Education

Algebra Use mental math to find an estimate of the unknown number.

10. $8,122 \div 9 = d$

d is about _____

11. $3,030 \div m = 600$

m is about _____

12. $4,883 \div 7 = h$

h is about _____

Problem Solving

Estimate. Check your estimates using multiplication.

13. Processes &Practices **Model Math** In August, 2,760 people attended concerts at Cooper Arena. About the same number of people went to see each of the 5 concerts. About how many people attended each concert?

_____

 Brain Builders

14. A mosaic at the art museum contains 2,889 tiles. Design a plan to divide the tiles into equal sections.

_____

_____

15. Mr. Morgan's ice-cream shop made $1,380 last weekend selling $2 scoops of ice cream. If he sells 100 more scoops this weekend, about how many scoops will he sell?

_____

Vocabulary Check

16. Circle the compatible numbers you could use to estimate $3,616 \div 9$.

$3,700 \div 9$ $3,600 \div 9$ $3,620 \div 9$

17. Test Practice Mrs. Scholl's helpers graded 632 tests during the school year. She had 3 helpers. About how many tests did each helper grade?

Ⓐ 315 tests   Ⓑ 310 tests   Ⓒ 210 tests   Ⓓ 200 tests

Name _____

Lesson 3
Hands On
Use Place Value to Divide

ESSENTIAL QUESTION
How does division affect numbers?

Build It

Find 39 ÷ 3.

 Model the dividend, 39.
Use the base-ten blocks to show
3 tens and 9 ones to show 39.

 Divide the tens.
The divisor is 3. So, divide the
tens into 3 equal groups.

There is _____ ten in each group.

 Divide the ones.
Divide the ones into 3 equal groups.

There are _____ ones in each group.
Draw a picture to show the equal groups.

There are _____ ten and _____ ones in each group.

$$3\overline{)39} \qquad \text{or} \qquad 39 \div 3 = \boxed{}\boxed{}$$

So, the quotient is _____ .

Some numbers do not divide evenly. The amount left over is called the **remainder**. Use the capital letter R to denote the remainder.

Find 68 ÷ 5 using base-ten blocks.

1 **Model the dividend.**
Use 6 tens and 8 ones to show 68.

2 **Divide the tens.**
Divide the tens into 5 equal groups. There is _____ ten in

each group. Regroup the ten that is left over into 10 ones.

There are _____ ones all together.

3 **Divide the ones.**
Divide the ones into 5 equal groups. Draw a picture to show the equal groups.

There is 1 ten and 3 ones in each group. There are 3 ones left over.

$$5\overline{)6\ 8}$$

You can also write this as
$68 \div 5 = 13\,R3$.

The ones left over are the remainder.

So, $68 \div 5 =$ _____ R _____.

1. **Processes &Practices** **2** **Reason** Explain what it means to have a remainder when dividing.

_____

_____

Practice It

Write the division sentence shown by each model.

2.

_____ ÷ _____ = _____

3.

_____ ÷ _____ = _____

Use models to find each quotient. Draw the equal groups.

4. 36 ÷ 2 = _____

There is _____ ten and _____
ones in each group.

The remainder is _____ .

5. 48 ÷ 3 = _____

There is _____ ten and _____
ones in each group.

The remainder is _____ .

6. 59 ÷ 4 = _____

There is _____ ten and _____ ones in each group.

The remainder is _____ .

Use models to solve.

7. There are 64 stickers. Each student gets 8 stickers. How many students are there?

_____

8. There are 73 party favors. Each bag can hold 9 party favors. How many full bags are there? How many party favors are left over?

_____

Processes &Practices **3** **Find the Error** A seal trainer has 47 seal treats. There are 4 seals. To find the number of seal treats each seal would receive, Mary drew the picture at the right to model 47 ÷ 4.

Look at Mary's drawing. Describe her error.

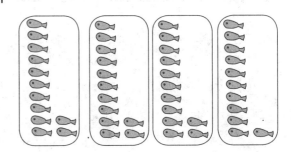

_____

_____

_____

Draw a correct picture to find 47 ÷ 4.

So, each seal will receive _____ treats.

How many seal treats are left over? _____

Write About It

10. How can place value help me divide?

_____

_____

_____

Name

Homework Helper

Need help? connectED.mcgraw-hill.com

Some numbers don't divide evenly. In this case, there is a remainder.

Find 43 ÷ 3.

1 Model the dividend, 43.

2 Divide the tens. The divisor is 3. So, divide the tens into 3 equal groups.

3 There is 1 ten left over. Regroup that ten as 10 ones.

4 10 ones plus the 3 ones you had to begin with makes 13 ones. Divide the 13 ones into four equal groups.

There are 1 ten and 4 ones in each group. There is 1 one left over. Left over ones are called the remainder. So, the quotient is 14 R1.

Practice

Write the division sentence shown by the model.

1.

Write the division sentence shown by the model.

2.

_____

 Problem Solving

Use models to find each quotient. Draw the equal groups.

3. There are 70 cards. Each person gets 5 cards. How many people are there?

$70 \div 5 =$ _____

There are _____ people.

4. **Processes &Practices** **Reason** There are 83 apples. Each bag can hold 4 apples. How many full bags are there? How many are left over?

$83 \div 4 =$ _____

There are _____ full bags.

There are _____ left over.

Vocabulary Check

5. Explain why, in division, there is sometimes a remainder.

_____

_____

_____

Name _____

Lesson 4
Problem-Solving Investigation
STRATEGY: Make a Model

ESSENTIAL QUESTION
How does division affect numbers?

Learn the Strategy

Ann's class bought 4 boxes of peaches at the orchard. There are 128 peaches in all. Each box has the same number of peaches. How many are in each box?

Understand

What facts do you know?

There are _____ peaches divided evenly into _____ boxes.

What do you need to find?

Find the number of _____ .

Plan

I will use base-ten blocks to model 128 ÷ 4.

Solve

Model 128. Divide the tens into four equal groups. Then divide the ones into four equal groups.

So, each box has _____ peaches.

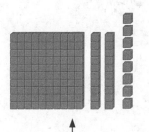

Since the hundred cannot be divided into 4 groups, trade 1 hundred for 10 tens.

Check

I can check by using repeated addition. 32 + 32 + 32 + 32 = 128

So, I know that my answer is reasonable.

Copyright © McGraw-Hill Education lynx/iconotec/Glowimages

Online Content at connectED.mcgraw-hill.com

Lesson 4 **347**

Caleb's family spent $420 on their road trip. Their road trip was 4 days long. If they spent the same amount of money each day, how much did they spend each day?

 Understand

What facts do you know?

_____

_____

What do you need to find?

_____

 Plan

_____

 Solve

4 Check

Does your answer make sense? Explain.

_____

Name

Apply the Strategy

Solve each problem by making a model.

1. Casey's mom is the baseball coach for his team. She spent $150 on baseballs. Each baseball cost $5. How many baseballs did she buy?

_____

2. Processes &Practices **2** **Reason** Each flowerpot costs $7. How many flowerpots can be bought with $285? Explain.

_____

_____

Brain Builders

3. Cindy spent $6 on a sandwich and smoothie. She spent $2 more on the sandwich than on the smoothie. How many smoothies could be bought with $124?

_____

4. Processes &Practices **5** **Use Math Tools** Use models to find the unknown numbers.

$254 \div \boxed{} = \boxed{}$ R4

5. Quincy found 120 shells during seven days at the beach. Is it possible that Quincy found the same number of shells each day? Explain.

_____

Review the Strategies

Use any strategy to solve each problem.
- Make a table.
- Choose an operation.
- Act it out.
- Draw a picture.

6. The calendar shows the number of days Carlota rides her bike each month. Each time she rides her bike, she travels 10 miles. Is it reasonable to say that Carlota will bike more than 500 miles in 6 months? Explain.

		September				
Sun	Mon	Tues	Wed	Thurs	Fri	Sat
					1	2 (B)
3 (B)	4	5	6	7	8 (B)	9
10 (B)	11	12	13 (B)	14	15	16 (B)
17	18	19	20	21	22 (B)	23
24 (B)	25	26	27	28 (B)	29	30 (B)

7. Processes &Practices **2 Reason** Paz and her scout troop make 325 granola bars for a fundraiser. Four granola bars are put in each bag. Paz says that there will not be any left over. Find and correct her mistake.

8. A coach ordered 6 soccer goals for $678. How much did each goal cost?

9. Gabriel has 268 miniature trains. He lines them up in 2 equal rows. How many trains are in each row?

MY Homework

Homework Helper

eHelp

Need help? connectED.mcgraw-hill.com

Daphne buys her mother a bunch of 12 flowers. Two of the flowers are daisies. She divides the remaining flowers in two groups. One group is tulips. How many of the flowers are tulips?

1 Understand

What do you know?

Daphne buys 12 flowers. Two are daisies. She divides the remaining flowers into two groups. One group is tulips.

What do you need to find?

I need to find how many flowers are tulips.

2 Plan

I will subtract the number of daisies. Then I will divide the remaining number of flowers by 2.

3 Solve

12 flowers − 2 daisies = 10 flowers

10 flowers ÷ 2 = 5 tulips

So, there are 5 tulips.

4 Check

I will use addition to check.

5 tulips + 5 other flowers + 2 daisies = 12 flowers

So, the answer is correct.

Problem Solving

Solve each problem by making a model.

1. There are two equal groups of 8 books on the shelf. One group has reading books. In the other group, three of the books are math books, and the rest are science books. How many science books are there?

_____

2. **Processes &Practices** 🔷 **Use Math Tools** There are 364 trees planted in 2 equal rows. One row has pine trees. The rest are oak trees. How many oak trees are there?

_____

3. There are 3 types of fish in a pond. There are a total of 162 fish. There is the same number of each type of fish. How many of each type of fish are in the pond?

_____

Brain Builders

4. Monica and her parents spent $170 on new swimsuits. Monica's swimsuit cost $35. Her mom's swimsuit cost twice as much as Monica's. How much less did her dad's swimsuit cost compared to her mom's?

_____

5. Joel divided 68 CDs evenly on 4 shelves in his bedroom. The CDs are arranged in alphabetical order, with the first shelf containing the letters A–E. If there are 13 CDs filed under A–D, how many CDs are filed under the letter E? Explain.

_____

Name ....

Lesson 5
Divide with Remainders

ESSENTIAL QUESTION
How does division affect numbers?

You have used models and fact families to divide. You can also use place value.

 Math in My World   Tools   Watch   Tutor

Example 1

Nolan and his family went to a water park during their vacation. Each seat on a water ride can hold 2 people. There are 39 people. How many seats will be needed?

Find 39 ÷ 2.

 Divide the tens.
How many groups of 2 are in 3 tens?
_____ group of ten

 Multiply, subtract, and compare.
Multiply.   2 × 1 = _____
Subtract.   3 − 2 = _____
Compare.   1 < 2

$$\begin{array}{r} 19 \text{ R1} \\ 2\overline{)39} \\ -2\downarrow \\ \hline 19 \\ -18 \\ \hline 1 \end{array}$$

 Bring down the ones.
Bring down 9 ones. There are now _____ ones.

 Divide the ones.
How many groups of 2 are in 19? _____ groups
Multiply.   2 × 9 = _____
Subtract.   19 − 18 = _____
Compare.   1 < 2

Check Use models to check.

19 seats are full. One additional seat has 1 person.

So, _____ seats are needed.

Example 2

Find 85 ÷ 3.

1 **Divide the tens.**

Divide How many groups
of 3 are in 8? 2 groups

Multiply 2 × 3 = 6.

Subtract 8 − 6 = 2

Compare. 2 < 3

Bring Down Bring down
the 5.

So, 85 ÷ 3 = _____ .

$$3\overline{)8\ 5}$$

2 **Divide the ones.**

Divide How many groups
of 3 are in 25? 8 groups

Multiply 8 × 3 = 24

Subtract 25 − 24 = 1

Compare. 1 < 3

Bring Down Since there
are no numbers to bring
down, 1 is the remainder.

Check

85 ÷ 3 = ☐ R ☐

☐ × 3 = 84   Then add the remainder.   84 + ☐ = 85

Talk MATH

When you divide
a number by 6, can
the remainder
be 6? Explain.

Guided Practice

Divide. Use multiplication to check.

1.   1 ☐ R ☐
$$2\overline{)2\quad 7}$$
− ☐
☐   7
− ☐
☐

Check: ☐ × 2 = ☐

☐ + ☐ = ☐

2.   ☐☐ R ☐
$$5\overline{)5\quad 9}$$
− ☐
☐   9
− ☐
☐

Check: ☐ × 5 = ☐

☐ + ☐ = ☐

Name _____

Independent Practice

Divide. Use multiplication to check.

3.
```
     1 ☐
  4)4  8
   -☐
  ─────
    ☐  8
    -☐
   ────
     ☐
```

4.
```
     1 ☐ R ☐
  5)5  3
   -☐
  ─────
    ☐  3
    -☐
   ────
     ☐
```

5.
```
    ☐☐ R ☐
  6)6  7
   -☐
  ─────
    ☐  7
    -☐
   ────
     ☐
```

6. 3)33

7. 3)73

8. 9)96

9. 69 ÷ 3 = _____

10. 77 ÷ 3 = _____

11. 99 ÷ 4 = _____

Algebra Use mental math to find the unknown.

12. $x \div 2 = 12$

13. $48 \div 4 = y$

14. $75 \div 5 = s$

$x =$ _____

$y =$ _____

$s =$ _____

Problem Solving

15. Processes &Practices **4** **Model Math** There are eight lions, four tigers, five cheetahs, six giraffes, seven hippos, and 78 monkeys at the City Zoo.

If each of the four zookeepers feeds the same number of animals, how many animals does each zookeeper feed? Explain.

_____

_____

16. Marlene makes $4 an hour babysitting. If she earned $48, how many hours did she babysit?

_____

17. Seven scouts need to sell 75 boxes of cookies. Each scout gets the same number of boxes. How many boxes will be left to sell?

_____

Brain Builders

18. Processes &Practices **1** **Keep Trying** Identify a two-digit dividend that will result in a quotient with a remainder of 1 when the divisor is 4.

_____

19. **Building on the Essential Question** Justin divided a number by 4 and got a remainder of 7. Explain how you know Justin's solution is not correct.

_____

_____

_____

Name ....

MY Homework

Homework Helper eHelp

Need help? connectED.mcgraw-hill.com

Find 62 ÷ 4.

1 Divide the tens.

$$4\overline{)62}$$ with 1 above

Divide. How many groups of 4 are in 6 tens? 1 group of ten.
Put 1 in the quotient over the tens place.

2 Multiply, subtract, and compare.

$$
\begin{array}{r}
1 \\
4\overline{)62} \\
-4 \\
\hline
2
\end{array}
$$

Multiply. $4 \times 1 = 4$
Subtract. $6 - 4 = 2$
Compare. $2 < 4$

3 Bring down the ones.

$$
\begin{array}{r}
1 \\
4\overline{)62} \\
-4 \downarrow \\
\hline
22
\end{array}
$$

Bring down 2 ones. There are 22 ones in all.

4 Divide the ones.

$$
\begin{array}{r}
15\ R2 \\
4\overline{)62} \\
-4 \\
\hline
22 \\
-20 \\
\hline
2
\end{array}
$$

How many groups of 4 are in 22? There are 5 ones in each group.
Put 5 in the quotient over the ones place.
Multiply. $4 \times 5 = 20$
Subtract. $22 - 20 = 2$
Compare. $2 < 4$

So, $62 \div 4 = 15\ R2$.

Practice

Divide. Check using multiplication.

1. 5)76

2. 2)39

3. 4)95

4. 6)86

5. 8)99

6. 3)80

Brain Builders

7. **Processes &Practices** **2** **Reason** Mitzi has 79 coins. She divides the coins equally among herself and her 3 brothers. Mitzi will keep any coins left over. How many coins does each person get?

_____

8. There are 26 third graders and 32 fourth graders going on a field trip. Each van can carry 10 students.

How many vans are needed? _____

How many students will be in each van?

_____

9. In Jonah's restaurant, each table seats 4 people. Jonah has 78 napkins to place at the tables.

How many tables can Jonah set with the napkins he has? _____

How many more napkins will Jonah need to complete

ten more tables? _____

10. **Test Practice** Wes bought 62 songs. He wants to put an equal number of songs on 5 different discs. How many songs will be left over?

   Ⓐ 4 songs   Ⓒ 2 songs

   Ⓑ 3 songs   Ⓓ 1 song

Lesson 6
Interpret Remainders

What should we do when there is a remainder?
We can interpret it in many ways.

Math in My World ▶ Watch 💬 Tutor

Example 1

Mandy wants to buy 4 books that each cost the same amount. If the total cost is $74, how much does each book cost?

Divide 74 by 4.

 Divide the tens.

Divide. There is 1 group of 4 in 7.

Multiply. 4 × 1 = 4

Subtract. 7 − 4 = 3

Compare. 3 < 4

Bring Down. Bring down the ones, 4.

$$\boxed{}\ \boxed{}\ R\ \boxed{}$$

$$4\overline{)7\quad 4}$$

2 **Divide the ones.**

Divide. There are 8 groups of 4 in 34.

Multiply. 8 × 4 = 32

Subtract. 34 − 32 = 2

Compare. 2 < 4

Bring Down. Since there are no numbers to bring down, the remainder is 2.

The remainder shows that each book will cost

a little more than _____ .

Example 2

Gracie has 64 prizes. She will put 3 prizes in each bag. How many bags will she have? Interpret the remainder.

1 Divide the tens.

Divide. How many groups of 3 are in 6? 2 groups

Multiply. $2 \times 3 = 6$

Subtract. $6 - 6 = 0$

Compare. $0 < 3$

Bring Down. Bring down the 4.

2 Divide the ones.

Divide. How many groups of 3 are in 4? 1 group

Multiply. $1 \times 3 = 3$

Subtract. $4 - 3 = 1$

Compare. $1 < 3$

Bring Down. Since there are no numbers to bring down, 1 is the remainder.

So, Gracie will have _____ bags. $64 \div 3 =$ _____ R _____

The remainder, _____, shows the number of prizes that Gracie will have left.

So, Gracie will have _____ prize left.

Check

_____ $\times 3 =$ _____ Multiply.

_____ $+ 1 =$ _____ Add the remainder.

Guided Practice

1. There are 45 people waiting for a bus. Each seat holds 2 people. How many seats will be needed? Divide. Interpret the remainder.

$45 \div 2 =$ _____ R _____ .

So, _____ seats will be needed.

Talk MATH

What kind of information can you get from a remainder?

Independent Practice

Divide. Interpret the remainder.

2. Gianna is at the school carnival. She has 58 tickets. It costs 3 tickets to play the basketball game. If she plays the basketball game as many times as she can, how many tickets will she have left?

   $3\overline{)58}$

   $58 \div 3 =$ _____

   So, there is _____ ticket left.

3. There are 75 people waiting in line to ride a roller coaster. Each car of the roller coaster holds 6 people. How many cars will be needed?

   $6\overline{)75}$

   $75 \div 6 =$ _____

   The answer is the next whole number, _____ .

   So, they will need _____ cars.

4. There are 4 cartons of orange juice in each package. If there are 79 cartons of orange juice, how many packages can be filled?

   $4\overline{)79}$

   $79 \div 4 =$ _____

   So, _____ packages can be filled.

5. The fourth grade classes are going on a field trip. There are 90 students in all. Each van can seat 8 students. How many vans will be needed?

   $8\overline{)90}$

   $90 \div 8 =$ _____

   The answer is the next whole number, _____ .

   So, they will need _____ vans.

Problem Solving

For Exercises 6 and 7, use the following information.

Parents are driving groups of children to the science center. Each van holds 5 children. There are 32 children in all.

6. **Processes &Practices** **Reason** How many vans are needed?

_____

_____

_____

7. Circle the true statement about the remainder.

   - You do not need to know anything about the remainder to solve this problem.

   - The remainder tells you that the answer is the next greatest whole number.

   - The remainder is the answer to the question.

Brain Builders

8. **Processes &Practices** **Use Number Sense** Brody is organizing his action figures for display. He wants to divide them equally among 9 displays. There are 78 action figures. Brody says that each display will have 6 figures. Find and correct his mistake.

_____

_____

9. **Building on the Essential Question** Why is it important to know how to interpret a remainder? Write a real-world problem and explain how to interpret the remainder.

_____

_____

_____

_____

Name ....

MY Homework

Homework Helper

Need help? connectED.mcgraw-hill.com

Miranda volunteers at the animal shelter. Each puppy gets 3 scoops of food per day. There are 50 scoops of food left in the bag. How many puppies will that amount feed for one day?

1 Divide the tens.

$$3\overline{)50} \quad \begin{array}{l} 1 \end{array}$$

How many groups of 3 are in 5?

There is 1 ten in each group. Put 1 in the quotient over the tens place.

Multiply, subtract, and compare.

$$\begin{array}{r} 1 \\ 3\overline{)50} \\ -\ 3\downarrow \\ \hline 20 \end{array}$$

Multiply. $3 \times 1 = 3$

Subtract. $5 - 3 = 2$

Compare. $2 < 3$, so bring down the ones to make 20.

2 Divide the ones.

$$\begin{array}{r} 16\ \text{R}\ 2 \\ 3\overline{)50} \\ -\ 3 \\ \hline 20 \\ -\ 18 \\ \hline 2 \end{array}$$

How many groups of 3 are in 20?

There are 6 ones in each group. Put 6 in the quotient over the ones place.

Multiply. $3 \times 6 = 18$

Subtract. $20 - 18 = 2$

Compare. $2 < 3$

There are no more numbers to bring down. The remainder is 2.

So, 50 scoops of food will feed 16 puppies for one day.

The remainder shows that there will be 2 scoops of food left over.

Practice

Divide. Interpret the remainder.

1. Tyler is planting 60 trees at the apple orchard. He will plant 8 trees in each row. How many full rows of trees will Tyler plant?

$$8\overline{)60}$$

_____ rows

2. Ms. Ling bought party hats for the 86 fourth graders at her school. The hats come in packages of 6. How many packages did Ms. Ling buy?

$$6\overline{)86}$$

_____ packages

Brain Builders

3. Wezi has sixty-three $5-bills. He takes the $5-bills to the bank to exchange them for $20-bills. How many $20-bills does Wezi get? Is there money left over? Explain.

_____

4. **Processes &Practices** **2** **Reason** Henry decorates the top of each cupcake with 3 walnuts. He has 5 packages that contain 27 walnuts each. Does Henry have enough to decorate 4 dozen cupcakes? Explain.

_____

_____

5. **Test Practice** Janice bought juice packets for the 15 players on the soccer team. The juice packets come in boxes of 6. How many boxes did Janice buy?

Ⓐ 5 boxes   Ⓑ 4 boxes   Ⓒ 3 boxes   Ⓓ 2 boxes

Check My Progress

Vocabulary Check

For Exercises 1–2, use pictures, words, or numbers to describe each vocabulary word.

1. remainder

2. compatible numbers

Concept Check

Complete each set of patterns.

3. $36 \div 6 =$ _____

    $360 \div 6 =$ _____

    $3,600 \div 6 =$ _____

4. $21 \div 7 =$ _____

    $210 \div 7 =$ _____

    $2,100 \div 7 =$ _____

Algebra Use mental math to estimate the unknown number.

5. $2,369 \div 6 = t$

    t is about _____

6. $\$6,285 \div y = \$700$

    y is about _____

7. $4,022 \div 8 = r$

    r is about _____

Divide. Use multiplication to check.

8. $3\overline{)63}$

9. $8\overline{)43}$

10. $3\overline{)79}$

Problem Solving

11. There are 4 members of a family planning a weekend camping trip. If the total cost of $200 is divided equally, how much will the trip cost for each person?

12. Nora earned 717 points on 9 tests. If she earned about the same number of points on each test, about how many points did she earn on each test?

Brain Builders

13. The fourth grade classes are going on a field trip. There are 85 students in all. Each van can seat 8 students. How many vans will be needed? Explain.

14. Ruby has 202 minutes left on her cell phone plan for the next 5 days. Is it possible for Ruby to use the same number of whole minutes each day? Explain.

15. **Test Practice** There are 39 paintbrushes. Each art bucket holds 6 paintbrushes. How many buckets will be needed to fit all of the paintbrushes?

   Ⓐ 3 buckets

   Ⓑ 5 buckets

   Ⓒ 7 buckets

   Ⓓ 9 buckets

Name

_____

Lesson 7
Place the First Digit

ESSENTIAL QUESTION
How does division affect numbers?

Sometimes the first digit of the dividend is less than the divisor. You may not be able to place the first digit of the quotient over the first digit of the dividend.

Math in My World

Example 1

Adriana's family went to Yellowstone National Park during their vacation. Suppose one of Yellowstone's geysers erupts every 7 minutes. How many times does the geyser erupt in 65 minutes?

Find $65 \div 7$.

Estimate $65 \div 7 \longrightarrow 70 \div 7 =$ _____

 Divide the tens.

$7\overline{)65}$ **Divide.** $6 \div 7$ Since $7 > 6$, you cannot divide the tens.

 Divide the ones.

$\boxed{}\ \text{R}\ \boxed{}$

$7\overline{)65}$
$\underline{-63}$
$\boxed{}$

Divide. There are 9 groups of 7 in 65.

Multiply. $7 \times 9 = 63$
Subtract. $65 - 63 = 2$

Compare. $2 < 7$
Bring Down. Since there are no more numbers to bring down, the remainder is 2.

So, the geyser will erupt 9 times in 65 minutes.

Check

9 is close to the estimate of 10. So, the answer is reasonable.

Copyright © McGraw-Hill Education

Example 2

A tennis coach has 125 tennis balls. There are 4 members on the team. How many balls does each player get for practice if each player gets the same number of balls?

Find 125 ÷ 4.

 Divide the hundreds.

Divide. 1 ÷ 4 Since 4 > 1, you cannot divide the hundreds.

 Divide the tens.

Divide. 12 ÷ 4 = 3 So, write 3 in the quotient over the tens place.
Multiply. 3 × 4 = 12
Subtract. 12 − 12 = 0
Compare. 0 < 4
Bring Down. Bring down the 5.

 Divide the ones.

Divide. There is 1 group of 4 in 5.
Multiply. 1 × 4 = 4
Subtract. 5 − 4 = 1
Compare. 1 < 4
Bring Down. Since there are no more numbers to bring down, the remainder is 1.

So, each team member gets _____ balls.

There will be _____ ball left over.

Estimation is one method that can be used to check division. Identify another method.

Guided Practice

Circle the correct place-value position to show where to place the first digit.

1. 2)33

tens

ones

2. 3)179

hundreds

tens

ones

Independent Practice

Divide. Use estimation to check.

3. 2)3 7

4. 5)49 R

5. 6)91 R

Estimate: _____   Estimate: _____   Estimate: _____

Divide. Use multiplication to check.

6. 4)7 9

7. 2)151 R

8. 3)286 R

Check: ____ × ____ = ____   Check: _____   Check: _____

____ + ____ = ____   _____   _____

Problem Solving

Every month, Americans throw out enough bottles and jars to fill up a giant skyscraper. All of these bottles and jars could be recycled.

9. When one aluminum can is recycled, enough energy is saved to run a television for 3 hours. How many cans need to be recycled to run a television for 75 hours?

_____

10. **Processes &Practices** **2** **Reason** Most Americans use 7 trees a year in products that are made from trees. How old is a person who has used 65 trees?

_____

11. **Processes &Practices** **1** **Plan Your Solution** When Kira's father's age, a, is divided by Kira's age, b, you get a quotient of 13 R1. Identify one possibility for their ages. Explain how you determined the ages.

$a =$ _____ $b =$ _____

_____

_____

12. **Processes &Practices** **3** **Find the Error** Colton is finding $53 \div 3$. Find and correct his mistake. How could Colton check his answer?

_____

_____

$$
\begin{array}{r}
1\ 1 \\
3\overline{)5\ 3} \\
-3\ \downarrow \\
\hline
3 \\
-3 \\
\hline
0
\end{array}
$$

13. **Building on the Essential Question** How do I know where to place the first digit of a quotient in a division problem?

_____

_____

Name

MY Homework

Homework Helper

Need help? connectED.mcgraw-hill.com

Find 145 ÷ 3.

 Divide the hundreds.

3)145

Sometimes it is not possible to divide the first digit of the dividend by the divisor. Since 3 is greater than 1, you cannot divide the hundreds.

Divide the tens.

$$\begin{array}{r} 4 \\ 3\overline{)145} \\ -12 \\ \hline 25 \end{array}$$

Divide. 14 ÷ 3 = 4

Multiply. 3 × 4 = 12

Subtract. 14 − 12 = 2

Compare. 2 < 3, so bring down the ones to make 25.

Divide the ones.

$$\begin{array}{r} 48 \text{ R1} \\ 3\overline{)145} \\ -12 \\ \hline 25 \\ -24 \\ \hline 1 \end{array}$$

Divide. 25 ÷ 3 = 8

Multiply. 3 × 8 = 24

Subtract. 25 − 24 = 1

Compare. 1 < 3

There are no more numbers to bring down; the remainder is 1.

Practice

Divide. Use estimation or multiplication to check.

R   R   R
1. 6)89   **2.** 6)73   **3.** 7)451

Divide. Use estimation or multiplication to check.

4. $3\overline{)105}$

5. $4\overline{)219}$ R

6. $8\overline{)254}$ R

7. $7\overline{)688}$ R

8. $5\overline{)396}$ R

9. $6\overline{)372}$

Brain Builders

10. **Processes &Practices** 5 **Use Math Tools** Chloe is making gift baskets with soaps and lotions. She has 127 soaps and 85 bottles of lotion. If Chloe puts 2 bottles of lotion and 3 soaps in each gift basket, what is the greatest number of baskets Chloe can make?

11. Dante has been putting one penny in his bank every day. Now he has $1.61. For how many weeks has Dante been adding pennies to his bank?

12. Nikki wants to buy a dog bed that costs $87. She earns $4 per week for folding laundry and $2 per week for pulling weeds. How many weeks will Nikki have to do her chores in order to earn enough money to pay for the dog bed? Explain.

13. **Test Practice** Ty's scout troop is picking up litter at the park. Each trash bag holds 8 pounds of litter. The scout troop collected a total of 230 pounds of litter. How many trash bags did they use?

Ⓐ 28   Ⓑ 29   Ⓒ 30   Ⓓ 31

Name ....

Lesson 8
Hands On
Distributive Property and Partial Quotients

You have used the Distributive Property to multiply. It can also help you divide.

Draw It Tools

Jake and his three brothers all went on separate vacations. They traveled 484 miles in all. They each traveled the same distance. How many miles did each person travel?

Find 484 ÷ 4.

1 Model 484 as (400 + 80 + 4).

400	80	4

2 Divide each section by 4. Write each quotient above the bar.

400	80	4

4

$400 \div 4 = \underline{\hspace{1cm}}$

$80 \div 4 = \underline{\hspace{1cm}}$

$4 \div 4 = \underline{\hspace{1cm}}$

3 Add the quotients. $\underline{\hspace{1cm}} + \underline{\hspace{1cm}} + \underline{\hspace{1cm}} = \underline{\hspace{1cm}}$

$484 \div 4 = \underline{\hspace{1cm}}$.

So, each person traveled $\underline{\hspace{1cm}}$ miles.

$484 \div 4 = \underline{\hspace{1cm}}$

$\underline{\hspace{1cm}} \times 4 = 484$

Check Use multiplication to check a division problem.

Partial quotients is a way to divide where you break the dividend into parts that are easier to divide.

Try It

There are 625 bars of soap. Each gift bag will have 5 bars of soap. How many gift bags can be made with the soaps?

Find 625 ÷ 5 using partial quotients.

1 **Divide the hundreds.**
500 is close to 625 and is compatible with 5.
Divide 500 by 5.

_____ is a partial quotient.

Subtract 500 from 625.

2 **Divide the tens.**
100 is close to 125 and is compatible with 5.
Divide 100 by 5.

_____ is a partial quotient.

Subtract 100 from 125.

Partial Quotients

```
    _____
  5)625      |  _____
  - 500      |
    125      |
  - 100      |  _____
     25      |
  -  25      |  _____
      0      |
```

3 **Divide the ones.**
Divide 25 by 5.

_____ is a partial quotient.

So, _____ gift bags can be made.

4 **Add the partial quotients.**

_____ + _____ + _____ = _____

625 ÷ 5 = _____

Talk About It

1. Draw an area model that could be used to find 346 ÷ 2 by using the Distributive Property.

2. **Processes &Practices** **Justify Conclusions** How are partial quotients similar to partial products?

_____

_____

_____

Name ....

Practice It

Divide. Use the Distributive Property. Complete the area models.

3. 624 ÷ 2

_____ + _____ + _____ = _____

624 ÷ 2 = _____

4. 848 ÷ 4

_____

848 ÷ 4 = _____

Divide. Use the Distributive Property. Draw area models.

5. 669 ÷ 3

6. 442 ÷ 2

_____

_____

_____

_____

Divide. Use the Distributive Property or partial quotients.

7. 7)826

8. 4)924

Apply It

**Use the Distributive Property or partial quotients
to solve Exercises 9–11.**

9. Blake's dad needs $165 for a new suitcase. If he saves an equal
   amount for 3 weeks, how much does he save each week?

10. There are 567 people at the basketball game. The bleachers are
    divided into 9 sections. There are an equal number of people in
    each section. How many people are in each section?

11. **Processes &Practices** 5 **Use Math Tools** Mrs. Schmitt baked 224 cookies
    in her bakery. She placed them in 2 rows in her display case.
    How many cookies are in each row?

12. In Exercise 10, why can the Distributive Property not be
    used to solve the problem?

13. **Processes &Practices** 6 **Explain to a Friend** When finding
    $180 \div 4$, is 40 or 60 a more reasonable partial quotient?
    Explain your answer to a friend.

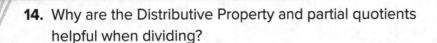

Write About It

14. Why are the Distributive Property and partial quotients
    helpful when dividing?

Name ....

MY Homework

Homework Helper eHelp

Need help? connectED.mcgraw-hill.com

Find 375 ÷ 5.

You can use the Distributive Property and an area model to divide.

 Model 375 as (300 + 70 + 5).

300	70	5

 Divide each section by 5.

	60	14	1
5	300	70	5

3 Add the partial quotients.

$$60 + 14 + 1 = 75$$
$$375 ÷ 5 = 75$$

So, 375 ÷ 5 = 75.

Check

Multiply to check your answer.

5 × 75 = 375, so the answer is correct.

Practice

Divide. Use the Distributive Property. Complete the area models.

1. 639 ÷ 3

_____ + _____ + _____ = _____

639 ÷ 3 = _____

2. 336 ÷ 6

_____ + _____ + _____ = _____

336 ÷ 6 = _____

Problem Solving

3. **Processes &Practices** **5** **Use Math Tools** A garden store has 868 plants. They are divided equally into two groups. How many plants are in each group?

_____

Divide. Use the Distributive Property or partial quotients.

4.
```
  3)762
  - 600        _____
  ─────
    162
  - 150        _____
  ─────
     12
  -  12        _____
  ─────
      0
```

_____ + _____ + _____ = _____

762 ÷ 3 = _____

5.
```
  2)426
  - 400        _____
  ─────
     26
  -  20        _____
  ─────
      6
  -   6        _____
  ─────
      0
```

_____ + _____ + _____ = _____

426 ÷ 2 = _____

Vocabulary Check

6. Explain why it can be helpful to use partial quotients when dividing.

_____

_____

_____

_____

_____

_____

Lesson 9
Divide Greater Numbers

ESSENTIAL QUESTION
How does division affect numbers?

Dividing three- and four-digit numbers is similar to dividing two-digit numbers.

 Math in My World Watch Tutor

Example 1

There are 678 people in line to ride a roller coaster. Each coaster car holds 6 people. How many coaster cars are needed so that everyone in line rides the coaster once?

Divide 678 by 6.

1 **Divide the hundreds.**

Divide. $6 \div 6 = 1$
Write 1 in the hundreds place.
Multiply. $6 \times 1 = 6$
Subtract. $6 - 6 = 0$
Compare. $0 < 6$
Bring down the tens.

$$\begin{array}{r} 1 \\ 6\overline{)678} \\ -\,6 \\ \hline 07 \end{array}$$

2 **Divide the tens.**

Divide. There is 1 group of 6 in 7.
Write 1 in the tens place.
Multiply. $6 \times 1 = 6$
Subtract. $7 - 6 = 1$
Compare. $1 < 6$
Bring down the ones.

$$\begin{array}{r} 113 \\ 6\overline{)678} \\ -\,6 \\ \hline 07 \\ -\,6 \\ \hline 18 \\ -18 \\ \hline 0 \end{array}$$

3 **Divide the ones.**

Divide. $18 \div 6 = 3$
Write 3 in the ones place.
Multiply. $6 \times 3 = 18$
Subtract. $18 - 18 = 0$
Compare. $0 < 6$

$678 \div 6 =$ _____

So, _____ coaster cars are needed.

Example 2

A roller coaster takes about 4 minutes to travel its 1,970-foot track. How many feet does the coaster travel in one minute?

Divide 1,970 by 4.

Estimate 1,970 ÷ 4 ⟶ 2,000 ÷ 4 = _____

1 **Divide the thousands.**

Since is 1 < 4, you cannot divide the thousands.

2 **Divide the hundreds.**

Divide. There are 4 groups of 4 in 19.
Multiply. Subtract. Compare. Bring down.

3 **Divide the tens.**

Divide. There are 9 groups of 4 in 37.
Multiply. Subtract. Compare. Bring down.

5 **Find the remainder.**

4 **Divide the ones.**

Divide. There are 2 groups of 4 in 10.
Multiply. Subtract. Compare. Bring down.

So, it travels a little more than _____ feet each minute.

Check The answer, a little more than _____, is close to the estimate of 500. So, the answer is reasonable.

Talk MATH

How would you mentally determine the number of digits in the quotient for 795 ÷ 5?

Guided Practice

Divide. Use estimation to check.

1. $2\overline{)286}$
$$\begin{array}{r} \square\square\square \\ 2\overline{)2\ 8\ 6} \\ -2 \\ \hline 0\ 8 \\ -8 \\ \hline 0\ 6 \\ -6 \\ \hline 0 \end{array}$$

2. $2\overline{)745}$

Independent Practice

Divide. Use estimation to check.

3. 2)324

4. 3)585

5. 2)1,573

Estimate:

Estimate:

Estimate:

Divide. Use multiplication to check.

6. 3)787

7. 2)849

8. 4)994

Check:

Check:

Check:

9. 3)1,863

10. 4)3,974

11. 4)2,611

Check:

Check:

Check:

Problem Solving

Use the following information for Exercises 12–13.

The White House is the official home and workplace of the President of the United States. President Theodore Roosevelt gave the White House its name, based on its color.

12. **Processes &Practices** **2** **Reason** It takes 570 gallons of paint to paint the outside of the White House. If the number of gallons used to paint each of its 4 sides is equal, how many gallons of paint are used on each side?

13. There are 132 rooms and 6 floors in the White House. If each floor has the same number of rooms, how many rooms would each floor have?

14. Britney reads a book in 9 days. If the book is 1,116 pages long, and she reads the same number of pages each day, how many pages does she read each day?

Brain Builders

15. **Processes &Practices** **1** **Make a Plan** Write a division problem that results in a quotient that is greater than 200 and less than 250. Explain how you chose your values.

16. **Building on the Essential Question** Do the quotients always have the same number of digits when dividing 3-digit numbers by 1-digit numbers? Explain.

Name ....

MY Homework

Homework Helper

eHelp

Need help? connectED.mcgraw-hill.com

Find 1,927 ÷ 4.

Estimate 1,927 is close to 2,000. 2,000 ÷ 4 = 500.

1 **Divide the thousands.**

4)1,927

Sometimes it is not possible to divide the first digit of the dividend by the divisor. Since 4 > 1, you cannot divide the thousands.

2 **Divide the hundreds.**

$$\begin{array}{r} 4 \\ 4\overline{)1,927} \\ -16 \\ \hline 32 \end{array}$$

Divide. 19 ÷ 4 = 4
Multiply. 4 × 4 = 16
Subtract. 19 − 16 = 3
Compare. 3 < 4
Bring down the 2.

3 **Divide the tens.**

$$\begin{array}{r} 48 \\ 4\overline{)1,927} \\ -16 \\ \hline 32 \\ -32 \\ \hline 07 \end{array}$$

Divide. 32 ÷ 4 = 8
Multiply. 4 × 8 = 32
Subtract. 32 − 32 = 0
Compare. 0 < 4
Bring down the 7.

4 **Divide the ones.**

$$\begin{array}{r} 481\text{R}3 \\ 4\overline{)1,927} \\ -16 \\ \hline 32 \\ -32 \\ \hline 07 \\ -4 \\ \hline 3 \end{array}$$

Divide. There is 1 group of 4 in 7.

Multiply. 4 × 1 = 4
Subtract. 7 − 4 = 3
Compare. 3 < 4

Since there are no more numbers to bring down, the remainder is 3.

1,927 ÷ 4 = 481 R3

Check

481 R3 is close to the estimate, 500, so the answer is reasonable.

Practice

Divide. Use estimation or multiplication to check.

1. 3)534

2. 7)2,761

3. 4)850

4. 8)1,074

5. 5)3,344

6. 6)5,244

Brain Builders

7. Ann has 9 days to read 203 pages of History and 211 pages in English. How many total pages should she read each day?

_____

8. **Processes &Practices** 5 **Use Math Tools** Marcel received an award for community service. The award included a check for $2,265. One business contributed $935. Two other businesses each contributed the same amount. How much did each of the other 2 businesses contribute? Explain your reasoning.

_____

9. **Test Practice** Eric collected 560 books to donate to a preschool. The books were divided equally among 5 classrooms. How many books did each classroom get?

Ⓐ 112 books   Ⓑ 110 books   Ⓒ 132 books   Ⓓ 512 books

Check My Progress

Vocabulary Check

1. Circle which of the following uses
   partial quotients to divide 362 by 2.

```
      181            181
   2)362          2)362      100
   - 2↓           - 200
     16             162       80
   - 16↓          - 160
     02              2       + 1
   -  2                      181
      0
```

Concept Check

Divide. Use estimation to check.

2. 4)56

3. 5)71 R

Estimate: _____   Estimate: _____

_____ ÷ _____ = _____   _____ ÷ _____ = _____

Divide. Use multiplication to check.

4. 3)345   **5.** 3)679

6. 697 ÷ 7 = _____   **7.** 883 ÷ 9 = _____

8. 917 ÷ 4 = _____   **9.** 775 ÷ 5 = _____

Problem Solving

10. There are 78 campers at a summer camp. There are 6 campers per cabin. How many cabins are there? _____

11. Carlo has $46 to spend on trading cards. If each pack of cards costs $3, how many packs of cards can he buy? _____

12. A tug-of-war team weighs a total of 774 pounds. The 6 members on the team each weigh the same amount. How much does each person weigh? _____

Brain Builders

13. A coach can purchase 9 hockey goals for $4,050, or 3 goals for $1,575. How much did each goal cost? Which option is the best deal? Explain.

_____

_____

14. Test Practice There are 4 times as many people at the city park as there are at the pool. If there are 456 people at the city park, how many people are at the pool?

   Ⓐ 111   Ⓒ 113

   Ⓑ 112   Ⓓ 114

Lesson 10
Quotients with Zeros

In division, a quotient will sometimes contain zeros.

 Math in My World Watch Tutor

Example 1

The Ramos family is going on a vacation to the behind-the-scenes tour of a wildlife reserve in a park. How much will it cost for each family member to go on the tour?

Find $327 \div 3$.

Cost of Tour	
Number of People	Cost ($)
3	327

1 Divide the hundreds.

$3 \div 3 = 1$

Multiply. Subtract. Compare. Bring down.

2 Divide the tens.

$2 < 3$

There are not enough tens to divide.
Write 0 in the tens place.
Multiply. Subtract. Compare. Bring down.

3 Divide the ones.

$27 \div 3 = 9$

Multiply. Subtract. Compare.

$327 \div 3 =$ _____.

So, it will cost each family member $ _____.

Example 2

The Kincaid family is going on vacation. They have to drive 415 miles to get to and from Dolphin Cove. How far is it to Dolphin Cove?

Divide 415 by 2.

1 **Divide the hundreds.**
$4 \div 2 = 2$
Multiply. Subtract. Compare. Bring down.

2 **Divide the tens.**
$1 < 2$. There are not enough tens to divide.
Write 0 in the tens place.
Multiply. Subtract. Compare. Bring down.

3 **Divide the ones.**
There are 7 groups of 2 in 15.
Multiply. Subtract. Compare.

4 **Find the remainder.**
A remainder of 1 tells you that the quotient is just over 207.

So, the distance to Dolphin Cove is a little more than _____ miles.

Check Use multiplication to check a division problem.

$415 \div 2 =$ _____ R _____

_____ $\times 2 = 414$ Then add the remainder. $414 +$ _____ $= 415$.

Guided Practice

Divide. Use multiplication to check.

1. $2\overline{)2\ 1\ 2}$
    $\underline{-\ 2}$
    $\ \ 0\ 1$
    $\underline{-\ \ 0}$
    $\ \ \ \ 1\ 2$
    $\underline{-\ 1\ 2}$
    $\ \ \ \ \ \ 0$

2. $2\overline{)6\ 1\ 7}$

Explain how to find the quotient of $624 \div 3$.

Name _____

Divide. Use multiplication to check.

3. 2)214

4. 3)327

5. 5)$545

6. $613 ÷ 3 = _____

7. 837 ÷ 4 = _____

8. 1,819 ÷ 2 = _____

Algebra Find the unknown.

9. 416 ÷ ■ = 208

10. 622 ÷ 3 = 207 R ■

11. $2,429 ÷ 3 = $■ R2

■ = _____

■ = _____

■ = _____

Problem Solving

Use the following information for Exercises 12 and 13.

Geocaching is an outdoor treasure hunting game in which participants use a Global Positioning System (GPS) to hide and seek "treasures" all over the world. The "treasures" are usually toys or trinkets.

12. Chad is saving his money to buy a GPS receiver so that he can go geocaching. He has 2 months to save $215. How much money does he need to save each month?

13. **Processes &Practices** **2** **Reason** Some of the treasures have been hidden on mountains. If the treasure is 325 feet away, how many yards away is it? (*Hint:* 3 feet = 1 yard)

Brain Builders

14. There are 408 students at a school. The cafeteria holds up to 90 students at a time. Are 4 lunch periods enough to allow every student to have lunch in the cafeteria? Explain.

15. **Processes &Practices** **1** **Keep Trying** Identify a 3-digit dividend that will result in a 3-digit quotient that has a zero in the tens place when the divisor is 6. Explain your reasoning.

16. **Building on the Essential Question** Why do I sometimes have to use 0 in a quotient?

Name

Homework Helper

Need help? connectED.mcgraw-hill.com

Find 614 ÷ 3.

 Divide the hundreds.

$6 \div 3 = 2$.

Multiply. Subtract. Compare. Bring down.

 Divide the tens.

$1 < 3$. There are not enough tens to divide.

Write 0 in the tens place.

Multiply. Subtract. Compare. Bring down.

 Divide the ones.

There are 4 groups of 3 in 14.

Multiply. Subtract. Compare.

 Find the remainder.

So, $614 \div 3 = 204$ R2.

```
        204 R2
   3)6 1 4
    − 6↓↓
        0 1
      −  0↓
          1 4
        − 1 2
            2
```

Practice

Divide.

1. 5)535

2. 4)826

3. 2)819

Divide.

4. $6\overline{)\$1,824}$

5. $7\overline{)3,517}$

6. $4\overline{)2,425}$

7. $3\overline{)626}$

8. $5\overline{)\$4,015}$

9. $8\overline{)1,613}$

Problem Solving

10. The water park had a total of 1,212 visitors on Friday, Saturday, and Sunday. If the same number of people visited each day, how many visitors were there on Sunday?

_____

Brain Builders

11. **Processes & Practices** **5** **Use Math Tools** The camping club spent $420 on four new tents. If the cost of each tent is evenly shared by 5 members, how much does each member pay per tent?

_____

12. Mallory has 535 e-mail messages in her Inbox. She sorts them evenly among folders labeled Family, Friends, Work, School, and Recipes. If each folder began with 25 messages, how many messages are in each folder now?

_____

13. **Test Practice** Mr. Lopez has collected 1,425 stamps. He sorts them into 7 piles. Which answer shows how many stamps are in each pile and how many are left over?

Ⓐ 204 R3

Ⓒ 220 R4

Ⓑ 229 R2

Ⓓ 203 R4

Lesson 11
Solve Multi-Step Word Problems

ESSENTIAL QUESTION ❓
How does division affect numbers?

Some word problems require more than one operation to solve.

 Math in My World Watch Tutor

Example 1

Jada's family bought memberships to the recreation center for a year. It cost **$532** for 4 regular memberships. A deluxe membership costs an additional **$35** per membership. How much does a deluxe membership cost?

Write an equation.

cost of each regular membership

add $35

deluxe membership

($532 ÷ 4) + $35 = ■ ← unknown

1 Divide.

☐ ☐ ☐

4)5 3 2
− ☐

☐ ☐
− ☐ ☐

☐ ☐
− ☐ ☐

☐

2 Add 35 to the quotient.

_____ + 35 = _____

The unknown is $ _____ .

Helpful Hint

Parentheses tell you which operations to perform first.

So, a deluxe membership costs $ _____ .

Example 2

Devin's family went snowboarding. Tickets cost $25 for kids and $30 for adults. Helmets cost $10 each. There are 3 kids and 2 adults in Devin's family. Everyone in Devin's family is getting a ticket and a helmet. Suppose they pay with $200. How much change will they receive?

Write an equation. The variable c can be used to represent the unknown.

 Find the total cost, c.

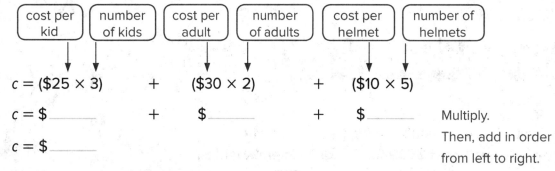

| cost per kid | number of kids | cost per adult | number of adults | cost per helmet | number of helmets |

$c = (\$25 \times 3)\quad + \quad (\$30 \times 2)\quad + \quad (\$10 \times 5)$

$c = \$\underline{\hspace{1.5cm}}\quad + \quad \$\underline{\hspace{1.5cm}}\quad + \quad \$\underline{\hspace{1.5cm}}$ Multiply.
Then, add in order
from left to right.

$c = \$\underline{\hspace{1.5cm}}$

The total cost is $\$\underline{\hspace{1.5cm}}$.

Talk MATH

What kinds of words help you decide which operations to use?

2 Find the amount of change.

$\$200 - \$185 = \$\underline{\hspace{1.5cm}}$.

So, Devin's family received $\$\underline{\hspace{1.5cm}}$ in change.

Guided Practice

1. The regular bookstore sold 345 books. The discount bookstore sold 3 times as many books. How many books were sold all together? Write an equation to solve the problem. Use a variable for the unknown.

$345 + (3 \times 345) = b$

$345 + \underline{\hspace{1.5cm}} = b$

$\underline{\hspace{1.5cm}} = b$

There are $\underline{\hspace{1.5cm}}$ books.

Name ....

Write an equation to solve each problem.
Use a variable for the unknown.

2. Ashlyn and her friends are making gingerbread houses. They need to divide 28 packages of gumdrops between 7 friends. There are 25 gumdrops in each package. How many gumdrops will each friend get?

Each friend gets _____ gumdrops.

3. Dominic ordered 210 pens. He divided them equally among his 10 friends. One of his friends, Benjamin, already had 27 pens. Then, Benjamin gave 13 of his pens to Peyton. How many pens does Benjamin have?

Benjamin has _____ pens.

4. Shannon is collecting art supplies. She has 48 crayons, 24 markers, and 16 stickers. She divided the crayons into 8 equal groups, the markers into 6 equal groups, and the stickers into 4 equal groups. She promised her brother that he could have one group from each type of art supplies. How many art supplies does her brother get?

Shannon's brother gets _____ art supplies.

Problem Solving

Use the following information to solve Exercises 5 and 6.

5. **Processes &Practices** **4** **Model Math** Hunter bought three boxes of cereal bars. Each box has 35 strawberry cereal bars. He also bought a box of 20 apple cereal bars and a box of blueberry cereal bars. There are a total of 150 cereal bars.

Write an equation to describe the number of cereal bars that Hunter bought. How many blueberry cereal bars did Hunter buy?

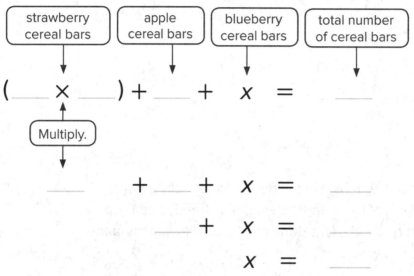

So, there are _____ blueberry cereal bars.

Brain Builders

6. **Processes &Practices** **2** **Reason** Hunter wants to divide 150 cereal bars among 7 people. How many cereal bars will Hunter have left? Give an example of how Hunter could divide 150 cereal bars among 4 friends and 3 brothers with none left over.

_____

_____

_____

7. **Building on the Essential Question** How can writing equations help me solve multi-step problems?

_____

_____

Name ....

MY Homework

Homework Helper

 eHelp

Need help? connectED.mcgraw-hill.com

Corinne has $45 saved from her allowance. She earns $7 per week. Corinne spends $2 of each allowance on a treat and saves the rest. How many weeks has Corinne been saving her money?

You need to find how much Corinne saves per week and divide $45 by that number.

1 Subtract.

$7 total weekly allowance

− $2 amount spent each week

$5 amount saved each week

2 Divide.

total saved $ saved each week

$45 ÷ $5 = 9 ← number of weeks

So, Corinne has been saving for 9 weeks.

Practice

1. Zoe needs to buy beads to make jewelry. For each earring, Zoe uses 3 small blue beads, 1 small metallic bead, and 2 large green beads. What is the total number of beads Zoe will need to buy to make 8 pairs of earrings?

_____

Problem Solving

Processes &Practices

2. **Reason** Miles has 327 action figures and 4 shoeboxes to store them in. Each shoebox holds 80 action figures. Does Miles have enough shoeboxes? If not, how many action figures are left over?

Brain Builders

3. Kyle has a notebook for each of his 13 classes. He puts 5 stickers on each notebook. There are 10 stickers on each sheet. How many sheets of stickers will Kyle need? Write an equation to solve. Use a variable for the unknown.

4. Rory is driving 584 miles to a family reunion. She will drive half of the distance by herself the first day. The second day, Rory and her cousin will share the driving equally. How many miles will her cousin drive? Write an equation to solve. Use a variable for the unknown.

5. **Test Practice** There are 278 houses in Hannah's neighborhood. She collected a total of $780 from her neighbors to donate to a local charity. Hannah received $5 at each house she went to. From how many houses did Hannah *not* collect money?

Ⓐ 156 houses   Ⓒ 55 houses

Ⓑ 122 houses   Ⓓ 125 houses

Copyright © McGraw-Hill Education

398 **Need more practice?** Download Extra Practice at connectED.mcgraw-hill.com

Fluency Practice

Multiply.

1.   429
   × 5

2.   357
   × 4

3.   189
   × 6

4.   672
   × 7

5.   2,416
   × 3

6.   7,515
   × 4

7.   4,219
   × 6

8.   5,413
   × 8

9.   3,035
   × 2

10.   8,107
   × 6

11.   4,050
   × 9

12.   8,063
   × 5

13.   83
   × 24

14.   27
   × 55

15.   64
   × 52

16.   92
   × 29

Fluency Practice

Divide.

1. 3)162

2. 5)261

3. 6)759

4. 4)529

5. 5)483

6. 4)244

7. 2)921

8. 8)327

9. 2)3,216

10. 6)4,842

11. 3)2,093

12. 5)3,526

13. 9)2,631

14. 3)5,111

15. 6)2,052

16. 4)1,729

Vocabulary Check

Use the word bank to match each example with its vocabulary word.

compatible numbers	equation	partial quotients
remainder	variable	

1. $122 \div 6 \longrightarrow 120 \div 6 = 20$

_____

2.
```
    23 R4
6)142   ↑
  -12
   22
  - 18
    4
```
_____

3. $80 \div 5 = 15$

_____

4. $360 \div 9 = x$
                    ↑
_____

5. _____

```
      125 R4
  6)754
   -600        100
    154         20
   -120
     34
    -30        + 5
      4        125
```

Concept Check

Divide. Use patterns and place value.

6. $600 \div 2 =$ _____

7. $7{,}200 \div 9 =$ _____

8. $6{,}400 \div 8 =$ _____

Estimate.

9. $715 \div 8$

10. $2{,}660 \div 9$

11. $8{,}099 \div 9$

Use the Distributive Property or partial quotients to divide.

12. $448 \div 2 =$ _____

13. $200 \div 8 =$ _____

Divide. Use multiplication to check.

14. $2 \overline{)64}$

15. $7 \overline{)694}$

16. $8 \overline{)783}$

17. $2 \overline{)2{,}157}$

18. $8 \overline{)487}$

19. $3 \overline{)451}$

Problem Solving

20. Clara's total score for 3 games of bowling is 312. If Clara earned the same score for each game, what was her score for each game?

_____

21. There are 3,250 buttons. If they are divided into groups of 8, how many groups are there? Interpret the remainder.

_____

_____

Brain Builders

22. The Nair family collected 2,400 pennies. The pennies will be divided evenly among the 4 children. How many dollars will each child get? How many dimes is that?

_____

23. On Saturday, 1,164 people saw a movie at Upcity Theater. On Sunday, 1,576 people saw the same movie. There were a total of 4 movie screens with the same number of people in each audience. About how many more people watched each screen on Sunday?

_____

24. Holden and Alma earned $32 by doing yard work in their neighborhood. They will share their money equally. If Holden saves $9, how much will he have to spend?

_____

25. **Test Practice** Casey and her two friends made $54 selling lemonade in their neighborhood. They will share their money equally. How much money will each person get?

Ⓐ $18   Ⓒ $27

Ⓑ $20   Ⓓ $54

Reflect

Use what you learned about division to
complete the graphic organizer.

Examples

**ESSENTIAL
QUESTION**

How does division
affect numbers?

Reflect on the **ESSENTIAL QUESTION** Write your answer below.

_____

_____

Performance Task

Hiking the Appalachian Trail

Debi is a college student. She plans to hike the length of the Appalachian Trail with 4 friends. The trail runs through 14 states, from Georgia to Maine.

Show all your work to receive full credit.

Part A

The friends plan to hike the trail over 4 summers, over a total of 8 months. They want to cover about the same distance each month. The trail is about 2,200 miles long. Use compatible numbers to estimate the number of miles they will hike each month. Tell whether your estimate is greater than or less than the actual number of miles they would hike each month.

Part B

Now estimate the distance the friends will hike each month by rounding the number of months and the distance to the greatest place value. Is this monthly distance less than, equal to, or greater than your estimate for Part A? Why do you think this is so?

Part C

Along the way, the hikers will reach the high point of the trip at Clingman's Dome on the North Carolina – Tennessee border. They will camp at an elevation of 3,256 feet. They will allow 10 hours to reach the top of the Dome at 6,643 feet. To reach the top, about how many feet in elevation must they gain each hour from their base camp to reach their goal? Show your work.

Part D

The hikers will carry backpacks with their own personal gear as well as group items. The weight of the group items will be shared equally among the hikers. The shared items weigh 145 pounds. Write an equation that can be used to find w, the weight of the shared items carried by each hiker. Then solve the equation.

Part E

Over the four years of hiking the Appalachian Trail, the hikers took 49 group pictures. Without dividing, tell if it is possible for the hikers to divide the pictures equally. Explain why or why not.

ESSENTIAL QUESTION

How are patterns used in mathematics?

Patterns In Our World

Watch a video!

Brain Builders

MY Chapter Project

Make a Game

1. In your group, discuss the different types of games you know, such as card games, spinner games, and trivia games. Make a list of these games below.

2. Discuss with your group what kind of game your math game will be. Remember, the game has to involve solving addition, subtraction, multiplication, and division equations.

3. On a separate sheet of paper, have someone in your group record the following about your game: your game's title, rules, and how to win. Work together to draw sketches of any boards, cards, or other items you'll use for your game.

4. After getting all the materials you need, work together to create your game. Make sure everyone in your group participates!

5. Play your game to test it out.

6. On game day, your group will play other groups' games. Your class can vote on which game they like best.

Name _____

Find each unknown.

1. $8 + \underline{\hspace{1cm}} = 11$

2. $\underline{\hspace{1cm}} + 5 = 9$

3. $6 + \underline{\hspace{1cm}} = 15$

4. $13 - \underline{\hspace{1cm}} = 7$

5. $\underline{\hspace{1cm}} - 4 = 8$

6. $18 - \underline{\hspace{1cm}} = 16$

7. Use the number sentence $12 + 15 + \blacksquare = 36$ to find how many books Tony read in August.

_____

Summer Reading Club	
Month	Number of Books Read
June	12
July	15
August	■

Find each value.

8. $8 + 1 + 6$

_____

9. $7 + 2 - 3$

_____

10. $2 + 10 - 6$

_____

11. $11 + 6 - 6$

_____

12. $12 - 3 + 4$

_____

13. $16 + 4 - 10$

_____

14. Each baseball uniform needs 3 buttons. Complete the table to find how many buttons are needed for 12 uniforms.

Uniforms	3	6	9	12
Buttons	9	18	27	

_____

Shade the boxes to show the problems you answered correctly.

How Did I Do?

1	2	3	4	5	6	7	8	9	10	11	12	13	14

Online Content at connectED.mcgraw-hill.com

MY Math Words

Vocab
abc

Review Vocabulary

equation   operations   unknown

Making Connections

Use the review vocabulary to describe each set of examples.
Then answer the question.

Algebraic Relationships

$4 \times 6 = 8 \times 3$

$81 \div 9 = 9$

$27 + 33$

$4,500 - 534$

48×7

$550 \div 50$

$\blacksquare \times 9 = 54$

$72 \div \blacksquare = 8$

How did you use patterns to categorize each set of examples?

_____

_____

_____

MY Vocabulary Cards

Lesson 7-5

input

Input (x)	Output (y)
2	9
4	11
6	13
8	15

$x + 7 = y$

Lesson 7-1

nonnumeric pattern

Lesson 7-2

numeric pattern

2, 4, 8, 16, 32, 64, 128

Lesson 7-5

output

Input (x)	Output (y)
2	9
4	11
6	13
8	15

$x + 7 = y$

Lesson 7-1

pattern

△ ◯ ▢ △ ◯ ▢ △ ◯ ▢

Lesson 7-2

rule

Rule: Multiply by 4	
Number of Squares	Number of Toothpicks
1	4
2	8
3	12

Lesson 7-3

sequence

2, 4, 8, 16, 32, 64, 128

sequence

Lesson 7-3

term

2, 4, 8, 16, 32, 64, 128

term

Ideas for Use

- Arrange cards in pairs. Explain your pairings to a friend.

- Write a tally mark on each card every time you read the word in this chapter or use it in your writing. Challenge yourself to use at least 5 tally marks for each card.

Patterns that do not use numbers.

Numeric means "of or relating to numbers." How does the prefix *non-* change the meaning of *numeric*?

_____

_____

A quantity that is changed to produce an output.

What is an antonym, or opposite-meaning word, for *input*?

_____

_____

The result of an input quantity being changed.

How would you explain the difference between *input* and *output* to another student?

_____

_____

_____

Patterns that use numbers.

Write a riddle to describe a numeric pattern. Challenge a friend to solve it.

_____

_____

_____

A statement that describes a relationship between numbers or objects.

Rule is a multiple-meaning word. Write a sentence using *rule* as a verb.

_____

_____

A list of numbers, figures, or symbols that follow a rule.

Describe a pattern you have seen on a shirt. How is this pattern similar to the math-related meaning of *pattern*?

_____

_____

_____

Each number in a numeric pattern.

Term is a multiple-meaning word. Write a sentence using another meaning of *term*.

_____

_____

The ordered arrangement of terms that make up a pattern.

Explain the word *sequence* in your own words.

_____

_____

MY Foldable

FOLDABLES® Follow the steps on the back to make your Foldable.

Input	Add 10	Subtract 3	Multiply by 3	Output Divide by 2
5	15	12	36	18
7	17	14	42	
9	19			
21	31	28		
33				

Output

Input

Name ....

Lesson 1
Nonnumeric Patterns

ESSENTIAL QUESTION ?
How are patterns used
in mathematics?

A **pattern** is a list of numbers, figures, or symbols
that follow a rule.

In this lesson, you will use and describe patterns that
do not use numbers. These are called **nonnumeric
patterns**. Many nonnumeric patterns are shape
patterns.

 Math in My World Watch Tutor

Example 1

**Edgar's room has a border of stars and moons. The stars and
moons show a repeating pattern. How many figures are in
this pattern unit? Copy and extend the pattern one time.**

 Find the pattern unit.

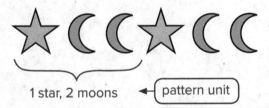

1 star, 2 moons ← pattern unit

Each pattern unit has _____ star and _____ moons.
The pattern repeats.

 Extend the pattern.

Copy the pattern as shown above. Then draw another star
and two moons.

So, there are _____ figures in this pattern unit.

Copyright © McGraw-Hill Education

Online Content at ⤏ connectED.mcgraw-hill.com

Patterns can also grow. They can get larger or smaller.

Example 2

Sal uses billiard balls to show a growing pattern. Use counters to model and describe the pattern. Then make an observation about the pattern.

Use counters to model the pattern.
Begin with one counter.

Each row adds one more billiard ball. So, place _____ counters in the second row and _____ counters in the third row.

Continue adding counters until you have 5 rows.

The pattern is add _____.

The number of billiard balls in each row alternates between odd and _____.

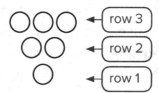

← row 3
← row 2
← row 1

Row	1	2	3	4	5
Number of Billiard Balls	1	2	3	4	5

Guided Practice

Extend each pattern. Draw the shapes on the lines.

1.

_____

2.

_____

Talk MATH

Make another observation about the pattern in Example 2.

Name ....

Independent Practice

Extend each pattern. Draw the shapes on the lines.

3. _____

4. _____

5. _____

6. _____

Draw eggs in the last carton to extend each pattern.

7.

8.

Problem Solving

Look for a Pattern Daniela notices a pattern in her scrapbook. The leaves on the first page are green. The second page has yellow leaves. The third page has green leaves. The next page has yellow leaves. Is this a repeating or a growing pattern? Explain.

_____

_____

Brain Builders

10. Processes &Practices 3 **Which One Doesn't Belong?** A growing pattern is shown at the right. Which figure below would not be in this pattern? Explain how you could change the figure so that it would belong in the pattern.

Figure 1   Figure 2   Figure 3

_____

_____

11. **Building on the Essential Question** Explain the difference between a nonnumeric growing pattern and a nonnumeric repeating pattern.

_____

_____

Provide one example of each type of pattern.

_____

_____

Name ....

Homework Helper

Need help? connectED.mcgraw-hill.com

Extend the pattern.

1 Find the pattern unit.

Each pattern unit has 1 blue pin, 1 red pin, and 1 yellow pin. This is a repeating pattern with 3 colors.

2 Extend the pattern.

Copy the pattern as shown above. Then draw another yellow pin.

Practice

Extend each pattern. Draw the shapes on the lines.

1.   _____

2.   _____

3.   _____

Identify or extend each pattern.

4. **Processes &Practices** **Model Math** Mrs. Arthur's fourth-grade class has art class every Monday. They go to computer class every Tuesday and Thursday. Mrs. Arthur's class has music on Wednesday and gym on Friday. Draw a picture to show their schedule for 2 weeks.

5. Mary Anne is planting flowers in her garden. She plants tulips in the first row, roses in the second row, and daisies in the third row. She repeats this pattern 4 times. Which kind of flower does she plant in the tenth row? Draw a picture to show Mary Anne's garden.

Vocabulary Check

Write a vocabulary term to complete each sentence.

nonnumeric patterns   patterns

6. _____ are lists of numbers, figures, or symbols that follow a rule.

7. Figures or symbols presented in repeating or growing designs are

_____ .

8. **Test Practice** Which comes next in the pattern?

Name ....

Lesson 2
Numeric Patterns

ESSENTIAL QUESTION
How are patterns used in mathematics?

Numeric patterns are patterns that use numbers.

A pattern is a list of numbers, figures, or symbols that follows a **rule**. A rule is a statement that describes a relationship between numbers or objects. You can find and extend a pattern by following a rule.

Math in My World

 Watch Tutor

Example 1

Carla sells 3 picture frames for $15 and 4 picture frames for $20. The pattern for the price of the frames is the same. How much money will Carla make if she sells 6 frames?

Identify and describe the pattern by dividing the total cost by the number of frames.

Carla sells 3 picture frames for $15, and 4 picture frames for $20.

$15 ÷ 3 = $5

$20 ÷ 4 = $5

So, one picture frame costs $5. The rule is to multiply the number of frames by $5.

Use the rule to extend the pattern.

Number of Frames	1	2	3	4	5	6
Cost	$5	$10	$15	$20	$25	

+$5   +$5   +$5   +$5   +$5

So, Carla will make 6 × $5, or _____, if she sells 6 picture frames.

Example 2

The chapters in Daniel's book follow a pattern. The even chapters have 12 pages. The odd chapters have 16 pages. The first page number of Chapter 1 is 1. Extend the pattern to find the page number of the last page in Chapter 6.

Daniel's Book	
Chapter	Ending
1	16
2	28
3	44
4	
5	
6	

+12
+16
+12
+16
+12

Extend the pattern.

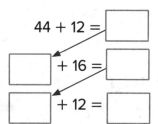

$44 + 12 =$ ☐

☐ $+ 16 =$ ☐

☐ $+ 12 =$ ☐

So, the last page number of Daniel's book is _____ .

Describe another pattern you see in this chart.

_____

_____

Talk MATH

Describe a real-world example of a growing numeric pattern.

Guided Practice

Identify, describe, and extend each pattern.

1. 9, 12, 15, 18, 21, _____

The pattern is add _____ .

2. 5, 6, 4, 5, 3, _____

The pattern is add _____ , then subtract _____ .

3. Antonio runs 30 minutes each day. Extend the pattern to find the number of minutes he will have run by Day 5.

Running Schedule					
Day	1	2	3	4	5
Time (min)	30				

Name ....

Identify, describe, and extend each pattern.

4. 3, 5, 7, 9, 11, _____

The pattern is _____.

5. 26, 30, 34, 38, 42, _____

The pattern is _____.

6. 8, 8, 6, 6, 4, _____

The pattern is _____

_____.

7. 28, 24, 28, 24, 28, _____

The pattern is _____

_____.

8. 10, 20, 30, 40, 50, _____

The pattern is _____.

9. 3, 6, 12, 15, 21, _____

The pattern is _____.

Find a rule and extend each pattern.

10. The table shows the number of puppets sold at a toy store. If the pattern continues, how many puppets will be sold on Day 5?

Puppets Sold					
Day	1	2	3	4	5
Puppets Sold	7	5	9	7	

The pattern is _____

_____.

So, _____ puppets will be sold on Day 5.

11. Ria uses blocks to build model towers. The table shows how many blocks she needs for different-sized towers. How many blocks will she need to build a 7-foot tower?

Towers	
Tower Height (ft)	Blocks Needed
4	128
5	160
6	192
7	

_____ blocks

Algebra Find the unknown in each pattern.

12. 24, 29, _____ , 39

13. 63, _____ , 47, 39

14. _____ , 17, 21, 25

15. _____ , 86, 82, 84, 80, 82, 78

Problem Solving

16. It is recommended that a person drink 64 fluid ounces of water each day. The pattern shows how many days it would take to drink 448 fluid ounces. Explain how another rule could be used to find the same answer.

Recommended Water in a Week							
Day	1	2	3	4	5	6	7
Amount (fl oz)	64	128	192	256	320	384	448

+64 + 64 + 64 +64 + 64 + 64

_____

_____

Brain Builders

Processes &Practices 8 **Look for a Pattern** The table at the right shows how much money Russell has each day.

Describe the rule for the pattern shown.

Russell's Budget	
Day	Russell's Money
Monday	$35
Tuesday	$31
Wednesday	$27
Thursday	$32
Friday	$28
Saturday	$24
Sunday	$29

_____

Overall, does Russell spend or earn more money? Explain.

_____

18. Processes &Practices 6 **Explain to a Friend** Create a number pattern involving two operations. Explain your pattern.

_____

_____

19. **Building on the Essential Question** Why is it important to look at more than just the first two numbers of a pattern to decide the rule for a pattern? Give an example of a pattern to explain your reasoning.

_____

_____

MY Homework

Homework Helper

Need help? connectED.mcgraw-hill.com

Identify, describe, and extend the pattern.

12, 7, 14, 9, 16, 11

1 Find the pattern by looking at how the numbers change. To get from 12 to 7, subtract 5. Then, to get from 7 to 14, add 7. See if the rule holds true for the rest of the numbers.

−5　+7　−5　+7　−5
12,　7,　14,　9,　16,　11

2 The rule "subtract 5, add 7" does hold true. Now extend the pattern.

+7　−5　+7　−5
12, 7, 14, 9, 16, 11,　18,　13,　20,　15

Practice

Identify, describe, and extend each pattern.

1. 39, 40, 36, 37, 33, 34, _____

The pattern is _____.

2. 64, 55, 46, 37, 28, 19, _____

The pattern is _____.

3. 53, 49, 52, 48, 51, 47, _____

The pattern is _____.

Problem Solving

Find a rule and extend each pattern.

4. **Processes &Practices** 2 **Use Number Sense** Duane and Mick are saving their money to build a tree house. Duane adds $5 to their piggy bank every other week. Mick adds $2 every week. So far they have saved $45. How many weeks have they been saving their money?

5. Janelle practices basketball every afternoon in her driveway. Each day, her goal is to make 4 more baskets than she made the day before. If she makes 10 baskets the first day and meets her goal for 2 weeks, how many baskets will Janelle make on the 14th day?

Brain Builders

6. For every 5 coins Carmen gets, she gives 2 to her brother Frankie. If Carmen has 9 coins, how many does Frankie have? Explain.

7. Kelli calls her grandmother every month. Every other month, Kelli also calls her cousin. If Kelli calls her cousin in January, how many calls will Kelli have made to her grandmother and her cousin by the end of August? Explain.

Vocabulary Check

Draw a line to match the word to its example.

8. rule • • 3, 7, 5, 9, 7, 11, 9

9. numeric pattern • • add 4, subtract 2

10. **Test Practice** Which pattern follows the rule "subtract 3, add 6"?

Ⓐ 18, 15, 21, 18, 24 Ⓒ 18, 15, 21, 18, 15

Ⓑ 18, 21, 15, 18, 12 Ⓓ 18, 24, 21, 27, 24

Lesson 3
Sequences

ESSENTIAL QUESTION
How are patterns used
in mathematics?

Patterns follow a rule. Each number in a numeric pattern is called a **term**. The ordered arrangement of terms that make up a pattern is called a **sequence**.

Math in My World

Example 1

Crystal starts reading her book on Monday. She reads 25 pages on the first day. Each day, she reads 25 pages. How many total pages will she have read by Tuesday, Wednesday, Thursday, and Friday?

The first term of the sequence is 25.

The rule is add 25.

Extend the pattern.

Day 1 Monday	Day 2 Tuesday	Day 3 Wednesday	Day 4 Thursday	Day 5 Friday
25	☐	☐	☐	☐

+25 +25 +25 +25

So, Crystal will have read _____ pages by Tuesday, _____ pages by Wednesday, _____ pages by Thursday, and _____ pages by Friday.

Example 2

The first term of a sequence is 65. The rule of the sequence is subtract 4. Find the next four terms in the sequence. Then make observations about the pattern.

 Find the next four terms.

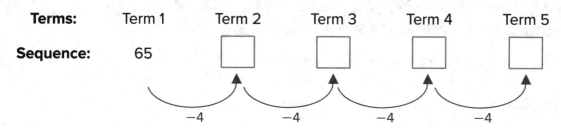

Terms:	Term 1	Term 2	Term 3	Term 4	Term 5
Sequence:	65	☐	☐	☐	☐

−4   −4   −4   −4

The next four terms in the sequence are ____, ____, ____, and ____.

Make observations about the pattern.

Circle whether the terms are all odd or even.　　　odd　　　even

Circle whether the terms increase or decrease.　　increase　　decrease

Extend the pattern to a total of 10 terms.

65, ____, ____, ____, ____, ____, ____, ____, ____, ____

Make another observation about the pattern.

The ones digits repeat the pattern 5, 1, ____, ____, and ____.

Guided Practice

Extend each pattern by four terms. Write an observation about the pattern.

Talk MATH

How does the operation of a rule affect the terms of a sequence?

1. Rule: add 7

Pattern: 8, ____, ____, ____, ____

Observation: _____

_____

2. Rule: subtract 10

Pattern: 90, ____, ____, ____, ____

Observation: _____

_____

Name ....

Independent Practice

Extend each pattern by four terms. Write an observation about the pattern.

3. Rule: add 9

Pattern: 7, _____ , _____ , _____ , _____

Observation: _____

_____

4. Rule: add 12

Pattern: 2, _____ , _____ , _____ , _____

Observation: _____

_____

5. Rule: subtract 9

Pattern: 87, _____ , _____ , _____ , _____

Observation: _____

_____

6. Rule: subtract 5

Pattern: 86, _____ , _____ , _____ , _____

Observation: _____

_____

7. Rule: multiply by 3

Pattern: 2, _____ , _____ , _____ , _____

Observation: _____

_____

8. Rule: multiply by 4

Pattern: 5, _____ , _____ , _____ , _____

Observation: _____

_____

9. Rule: divide by 2

Pattern: 64, _____ , _____ , _____ , _____

Observation: _____

_____

10. Rule: divide by 5

Pattern: 625, _____ , _____ , _____ , _____

Observation: _____

_____

11. Refer to the sequence 11, 16, 21, 26, 31, 36. Explain why the terms in the sequence will continue to alternate between even and odd numbers.

_____

_____

_____

Problem Solving

12. Each pumpkin costs $8. Jaime has already bought $24 worth of pumpkins. Suppose he buys five more pumpkins. How much will he spend in all after he buys each pumpkin? Write a sequence.

13. **Processes &Practices** ③ **Draw a Conclusion** The rule of a pattern is multiply by 3. The first term is 7. What are the next five terms in the sequence?

Write two observations you can make about the pattern.

Brain Builders

14. **Processes &Practices** ② **Use Number Sense** Write a sequence with at least 5 terms that forms a pattern. Identify the rule.

Apply your rule to a new starting value to produce a new sequence.

15. **Processes &Practices** ② **Reason** The first term in a sequence is an odd number. The rule is to multiply by 2. Are the rest of the terms in the sequence even, odd, or both? Explain.

16. **Building on the Essential Question** How can I find patterns?

Copyright © McGraw-Hill Education

Name ....

MY Homework

Homework Helper

Need help? connectED.mcgraw-hill.com

**Extend the pattern described below by four terms.
Then, note two observations about the pattern.**

Use repeated subtraction to extend the pattern.

First Term: 46

$$
\begin{array}{cccc}
46 & 39 & 32 & 25 \\
-7 & -7 & -7 & -7 \\
\hline
39 & 32 & 25 & 18
\end{array}
$$

Rule: Subtract 7

So, the sequence is 46, 39, 32, 25, and 18.
The terms in the sequence decrease. The terms in the
sequence also alternate between even and odd numbers.

Practice

Extend each pattern by four terms. Write an observation about the pattern.

1. Rule: add 8

Pattern: 5, _____ , _____ , _____ , _____

Observation: _____

2. Rule: multiply by 2

Pattern: 3, _____ , _____ , _____ , _____

Observation: _____

3. Rule: subtract 20

Pattern: 175, _____ , _____ , _____ , _____

Observation: _____

4. Extend the pattern below by four terms. Write an observation about the pattern.

Rule: multiply by 10

Pattern: 26, _____, _____, _____, _____

Observation: _____

Brain Builders

Processes &Practices **8**   **Look for a Pattern** Brad puts an equal amount of money in his savings account once a month. He started with $27. The next month, he had $44 in his account. Two months after that, he had $78 in his account. How much money will Brad have in his account after 6 months? Describe a rule. Then solve.

_____

6. On Monday, a toy store sold 4 race cars. On Tuesday, it sold 8 race cars. On Wednesday, it sold 16 race cars. Suppose this pattern continues. How many more race cars will be sold on Friday than on Thursday? Describe a rule. Then solve.

_____

_____

Vocabulary Check   [Vocab abc]

Write a vocabulary word to complete each sentence.

sequence   term

7. Each number in a numeric pattern is a _____.

8. A _____ is the ordered arrangement of terms that make up a pattern.

9. **Test Practice**   Identify the next term in the sequence.   171, 141, 151, 121, 131, _____

Ⓐ 161   Ⓑ 141   Ⓒ 121   Ⓓ 101

Name

Lesson 4
Problem-Solving Investigation
STRATEGY: Look for a Pattern

ESSENTIAL QUESTION
How are patterns used in mathematics?

Learn the Strategy

Daniel is training for a walk-a-thon. In the first week, he walked a total of five miles. In the second week, he walked a total of 7 miles. In the third week, he walked a total of 9 miles. Based on his pattern, how many miles will he walk in the fourth week?

1 Understand

What facts do you know?

Daniel walked _____ miles the first week, _____ miles the second week, and _____ miles the third week.

What do you need to find?

the number of miles Daniel will walk the _____ week

2 Plan

I will look for a pattern to solve the problem.

3 Solve

The sequence of the pattern is: 5, 7, 9.

The rule to the pattern is _____ .

Based on the rule, the next term in the sequence is _____ .

So, Daniel will walk _____ miles during the fourth week.

4 Check

Does your answer make sense? Explain.

Taryn made 15 hair ribbons on Monday, 21 hair ribbons on Tuesday, and 27 hair ribbons on Wednesday. Based on her pattern, how many hair ribbons will she make on Thursday?

1 Understand

What facts do you know?

_____

_____

What do you need to find?

_____

2 Plan

_____

3 Solve

4 Check

Does your answer make sense? Explain.

_____

_____

Name ....

Solve each problem by looking for a pattern.

1. **Processes &Practices** **Look for a Pattern** A store sold 48 model airplanes in August, 58 model airplanes in September, and 68 model airplanes in October. Suppose this pattern continues. How many model airplanes will be sold in December?

Brain Builders

2. The table shows how many tickets were sold for the school play each day.

Day	Number of Tickets
Monday	312
Tuesday	316
Wednesday	320
Thursday	324

Based on the pattern, how many tickets will be sold on Saturday?

3. There are 80 picnic tables at the park. During the first weekend of the summer, there were 40 available tables. During the second weekend, there were 20 available tables. During the third weekend, there were 10 available tables. Based on the pattern, how many picnic tables will be available during the fourth weekend of the summer? Explain.

Review Strategies

4. **Processes &Practices** **2** **Use Symbols** A theater can hold 200 people. Two groups rented out the theater. The first group has 92 people and the other has 107 people. Are there enough seats for everyone? Use symbols to explain.

_____

5. In one hour, Frank earns the money shown below. How much does he earn in 7 weeks if he works 3 hours each week?

_____

6. What is the next number in the pattern 2, 5, 11, 23, ▇?

_____

7. Katie sold 153 purses at a craft fair. How much money did she earn if each purse cost the amount shown below?

_____

MY Homework

Homework Helper eHelp

Need help? connectED.mcgraw-hill.com

Solve the problem by looking for a pattern.

At ShopSmart, customers receive a $3 coupon if they spend $20, a $6 coupon if they spend $40, and a $9 coupon if they spend $60. If a customer spends $80, how many dollars worth of coupons will he or she receive if the pattern continues?

1 Understand

What facts do you know?

Customers get a $3 coupon for every $20 they spend.

What do you need to find?

I need to find how many dollars in coupons a customer gets for spending $80.

2 Plan

I will find a pattern to solve the problem.

3 Solve

The sequence of the pattern of the coupons is $3, $6, and $9.

The rule for the pattern is +$3.

Based on the rule, the next term in the sequence is $12.

So, customers receive a $12 coupon for spending $80.

4 Check

Does the answer make sense?
$9 + $3 = $12, so the answer makes sense.

Problem Solving

Solve each problem by looking for a pattern.

1. **Processes &Practices** ① **Plan Your Solution** Angela opened a new bakery. She got orders for 2 cakes the first week, 4 cakes the second week, and 8 cakes the third week. If the pattern continues, how many cake orders will Angela get the fourth week?

2. Manuel saw the following birds this week: 2 blue jays on Monday, 5 cardinals on Tuesday, 4 blue jays on Wednesday, 7 cardinals on Thursday, and 6 blue jays on Friday. If the pattern continues, what is the number and type of bird Manuel will see Saturday?

Brain Builders

3. The house numbers on the north side of Flynn Street are even. In one block, the house numbers begin 1022, 1032, 1042, and 1052. If Taylor is at 1022 Flynn Street, how many houses away is 1082?

4. Jessica is voting for dancers in a contest. She votes for the first contestant 5 times, the second contestant 9 times, and the third contestant 13 times. If she continues this pattern, how many total votes will Jessica cast for the first four contestants? Explain.

5. A towel has a repeating pattern of 2 green stripes, then 3 blue stripes, and then 1 yellow stripe. If the towel has 20 stripes all together, how many green stripes are on it? Explain.

Check My Progress

Vocabulary Check

Identify each pattern as a **nonnumeric pattern** or a **numeric pattern.**

1.

_____

2. 43, 46, 47, 50, 51, 54, 55

_____

3. 98, 88, 78, 68, 58, 48

_____

4.

_____

Use the pattern below for Exercises 5–7.

2, 6, 18, 54, 162

5. Put a circle around one **term** in the pattern.

6. Underline the **sequence**.

7. Write the **rule** for this pattern. _____

Concept Check

8. Extend the pattern. Draw the shapes on the lines.

 _____

9. Extend the pattern. Write the rule.

3, 8, 13, 18, 23, _____   The rule is _____ .

10. Extend the pattern below by four terms. Write an observation about the pattern.

Rule: subtract 6

Pattern: 76, _____ , _____ , _____ , _____

Observation:

_____

_____

Problem Solving

11. On Mondays, Wednesdays, and Fridays, Luke packs his lunch. On Tuesdays and Thursdays, Luke buys his lunch. Draw a nonnumeric pattern to show the pattern of Luke's lunch for two weeks.

_____

 Brain Builders

12. Bob swims 10 laps on even numbered dates. He swims 15 laps on odd numbered dates. How many laps has he completed by the sixth of the month? Describe a pattern to solve the problem.

_____

13. Test Practice A nonnumeric pattern is shown below.

Which shows the next three objects in the pattern?

Ⓐ

Ⓒ

Ⓑ

Ⓓ

Name
....

Lesson 5
Addition and Subtraction Rules

ESSENTIAL QUESTION
How are patterns used in mathematics?

You can use a rule to write an equation that describes a pattern between **input** and **output** numbers. Tables can be used to show how input numbers change in the same way each time, creating a new output number.

 Math in My World

Example 1

Mr. Mathis is creating a table to show how input numbers are changed. Write an equation that describes the pattern in the table. Complete the table.

Input (x)	Output (y)
2	9
4	11
6	13
8	
10	
12	

Pattern: $2 + \underline{\hphantom{x}} = 9$

$4 + \underline{\hphantom{x}} = 11$

$6 + \underline{\hphantom{x}} = 13$

Rule: Add _____ .

Equation: $x + \underline{\hphantom{x}} = y$

↑ Input   ↑ Output

Use the rule to complete the table.

So, the equation that describes the pattern is _____ .

Example 2

A pizza shop offers $3 off any order over $10. Use the rule and equation to find the next four output numbers.

Input (c)	Output (d)
$11	$8
$12	
$14	
$16	
$18	

Rule: Subtract 3.

Equation: $c - \$3 = d$

Input ↑   Output ↑

Find the next four numbers when the input c is $12, $14, $16, and $18.

$c - \$3 = d$	$c - \$3 = d$	$c - \$3 = d$	$c - \$3 = d$
$\$12 - \$3 = \$\boxed{}$	$\$14 - \$3 = \$\boxed{}$	$\$16 - \$3 = \$\boxed{}$	$\$18 - \$3 = \$\boxed{}$

So, the next four amounts are _____ .

Describe another pattern you see in this chart.

_____

_____

Guided Practice

1. Write an equation that describes the pattern.
Then use the equation to find the next three output numbers.

Input (a)	5	9	13	17	21	25
Output (b)	9	13	17			

_____

Explain what you should do if you test a number in an equation and it does not work.

Independent Practice

Write an equation that describes the pattern. Then use the equation to find the next two output numbers.

2.

Input (m)	11	16	21	26	31
Output (n)	2	7	12		

3.

Input (s)	2	6	10	14	18
Output (t)	15	19	23		

Equation: _____

Equation: _____

Use the rule to find the next four output numbers.

4.

Rule: $f + 3 = h$	
Input (f)	Output (h)
3	6
6	
9	
12	
15	

5.

Rule: $v - 11 = w$	
Input (v)	Output (w)
16	5
22	
28	
34	
40	

6.

Rule: $g - 5 = h$	
Input (g)	Output (h)
14	9
19	
24	
29	
34	

Create an input/output table for each equation.

7. $y + 4 = z$

8. $a - 7 = c$

9. Describe a pattern you see in Exercise 2.

Problem Solving

The table shows what a taxi company charges
in dollars *c* for every *m* miles traveled.

Taxi Rates	
Input (*m*)	Output (*c*)
10	$12
15	$17
20	$22
25	▢
30	▢

Processes
10. **&Practices** 2 **Use Algebra** Use the table to write an
equation for this situation.

11. Find the costs of a 25-mile trip and a 30-mile trip.

12. Use the equation you wrote for Exercise 10 to find
the cost of a 60-mile trip.

13. A different taxi company uses the equation
$c = m + \$4$ to determine their charges. Find the
cost of a 15-mile trip.

Brain Builders

Processes
14. **&Practices** 4 **Model Math** Write a real-world problem that
can be represented by the table. Include the solution to your problem.

Input (*h*)	1	2	3	4	5
Output (*m*)	$10	$20	$30	▢	▢

15. **Building on the Essential Question** How can I find the rule of
a pattern?

Name

Homework Helper

Need help? connectED.mcgraw-hill.com

**Write an equation that describes the pattern in the table.
Then use the equation to find the next three output numbers.**

Input (d)	12	15	18	21	24	27
Output (f)	19	22	25	■	■	■

The rule is add 7. The letter *d* represents the input, and the
letter *f* represents the output. So, the equation is $d + 7 = f$.

Use the equation to find the next three output numbers:

$$21 + 7 = 28$$
$$24 + 7 = 31$$
$$27 + 7 = 34$$

So, the completed table looks like this:

Input (d)	12	15	18	21	24	27
Output (f)	19	22	25	28	31	34

Practice

**Write an equation that describes the pattern. Then use the equation to
find the next three output numbers.**

1.

Input (a)	Output (b)
$2	$27
$4	$29
$6	
$8	
$10	

Equation: _____

2.

Input (s)	Output (t)
87	76
80	69
73	
66	
59	

Equation: _____

Write an equation that describes the pattern. Then use the
equation to find the next three output numbers.

3.

Input (x)	Output (y)
22	17
26	21
30	
34	
38	

Equation: _____

4.

Input (c)	Output (d)
0	8
5	13
10	
15	
20	

Equation: _____

Brain Builders

Jeremy's class is going on a field trip. The school will bring
all the students who are there that day plus 4 chaperones.

5. Write an equation for this situation. Explain how your
equation relates to the given information.

_____

_____

Input (s)	Output (p)

6. **Processes &Practices** **Use Math Tools** Complete the table
to show how many people will go if there are 25, 27, 29, 31,
or 33 students. Explain how you can check your answer.

_____

_____

Vocabulary Check

Draw a line to match each word to its meaning.

7. input

• a number before an operation is performed

8. output

• a number that is the result of an operation

9. **Test Practice** Refer to the equation $a - 6 = b$. If $a = 45$, what is the value of b?

Ⓐ 16   Ⓑ 39   Ⓒ 51   Ⓓ 60

Name ....

Lesson 6
Multiplication and Division Rules

ESSENTIAL QUESTION
How are patterns used in mathematics?

You can write a multiplication or division equation to extend a pattern.

Math in My World

Example 1

Charles washes cars to earn money. If he washes 2 cars, he earns $12. If he washes 6 cars, he earns $36. Write an equation to describe the pattern. Then use the equation to find how much money Charles will earn if he washes 8, 10, and 12 cars.

Complete the table. Then look for the pattern that describes a rule.

Pattern: 2 × _____ = 12

                4 × _____ = 24

                6 × _____ = 36

Rule: Multiply by _____.

Equation: $a ×$ _____ $= b$

Cars Washed	Amount Earned ($)
Input (a)	Output (b)
2	12
4	24
6	36
8	
10	
12	

Find the next three output numbers when the input a is 8, 10, and 12.

$a × 6 = b$ $a × 6 = b$ $a × 6 = b$

$8 × 6 =$ _____ $10 × 6 =$ _____ $12 × 6 =$ _____

So, Charles will earn $ _____ , $ _____ , and $ _____ .

Example 2

It costs $4 for each box of crackers. The equation is shown below. Use the equation to complete the table.

Rule: Divide by 4.

Equation: $g \div 4 = h$

Input ⟶   ⟵ Output

Find the next five output numbers when the input g is 8, 12, 16, 20, and 24.

Total Cost ($)	Boxes of Crackers
Input (g)	Output (h)
4	1
8	
12	
16	
20	
24	

$g \div 4 = h$ $g \div 4 = h$ $g \div 4 = h$ $g \div 4 = h$ $g \div 4 = h$

$8 \div 4 =$_____ $12 \div 4 =$_____ $16 \div 4 =$_____ $20 \div 4 =$_____ $24 \div 4 =$_____

So, the next five output numbers are _____.

Describe another pattern you see in this table.

_____

_____

Guided Practice

1. Write an equation that describes the pattern. Then use the equation to find the next three output numbers.

Input (w)	2	4	6	8	10	12
Output (v)	12	24	36			

Equation: _____

Describe a pattern you see in this chart.

_____

_____

How are a rule and an equation alike? How are they different?

Independent Practice

Write an equation that describes the pattern. Then use the equation to find the next three output numbers.

2.

Input (m)	1	3	5	7	9	11
Output (n)	5	15	25			

Equation: _____

3.

Input (b)	2	4	6	8	10	12
Output (c)	14	28	42			

Equation: _____

4.

Input (j)	4	8	12	16	20	24
Output (k)	1	2	3			

Equation: _____

5.

Input (e)	10	20	30	40	50	60
Output (f)	2	4	6			

Equation: _____

6.

Input (x)	16	24	32	40	48	56
Output (y)	2	3	4			

Equation: _____

7.

Input (t)	12	10	8	6	4	2
Output (v)	24	20	16			

Equation: _____

Create an input/output table for each equation.

8. $a \times 5 = b$

9. $c \div 6 = d$

10. Describe a pattern you see in Exercise 6.

_____

_____

Problem Solving

Sari makes bead necklaces. The table shows the number of blue beads and green beads Sari uses.

Blue Beads	Green Beads
Input (j)	Output (k)
3	1
9	3
15	5
21	▨
27	▨
33	▨

11. **Processes &Practices** ➋ **Use Algebra** Write an equation that describes the relationship between green beads and blue beads.

_____

12. How many green beads does Sari need if she is using 36 blue beads?

_____

Brain Builders

13. How many beads does Sari have in all if she has 30 green beads? Explain.

_____

14. **Processes &Practices** ➋ **Reason** Circle the operation that can be used to write an equation for the input/output table to the right. Explain.

Input (m)	Output (n)
1	2
2	4
3	6

addition   subtraction   multiplication   division

_____

_____

15. **Building on the Essential Question** How can an input/output table help me solve a real-world problem?

_____

_____

Name _____

MY Homework

Homework Helper

Need help? connectED.mcgraw-hill.com

Write an equation that describes the pattern in the table below. Then use the equation to find the next three output numbers.

Input (k)	2	4	6	8	10	12
Output (m)	6	12	18			

The rule is multiply by 3. The letter k represents the input, and the letter m represents the output. So, the equation is $k \times 3 = m$.

Use the equation to find the next three output numbers:

$8 \times 3 = 24$
$10 \times 3 = 30$
$12 \times 3 = 36$

So, the completed table looks like this:

Input (k)	2	4	6	8	10	12
Output (m)	6	12	18	24	30	36

Practice

Write an equation that describes the pattern. Then use the equation to find the next three output numbers.

1.

Input (a)	Output (b)
7	1
14	2
21	
28	
35	

Equation: _____

2.

Input (s)	Output (f)
99	33
84	28
69	
54	
39	

Equation: _____

Write an equation that describes the pattern. Then use the equation to find the next three output numbers.

3.

Input (x)	Output (y)
$5	$40
$6	$48
$7	
$8	
$9	

Equation: _____

4.

Input (c)	Output (d)
50	10
45	9
40	
35	
30	

Equation: _____

Problem Solving

Shawna found out there are 4 yellow pencils for every blue pencil.

5. **Processes &Practices** **2** **Use Algebra** Write an equation for this situation. Explain your equation.

_____

_____

_____

Brain Builders

6. Complete the table to show how many yellow pencils there are if there are 5, 7, 9, 11, or 13 blue pencils. How many of each type of pencil does Shawna have if she has 55 pencils?

Input (s)	Output (p)

7. **Test Practice** Refer to the equation $a \times 9 = b$. If $a = 3$, what is the value of b?

Ⓐ 3   Ⓑ 12   Ⓒ 18   Ⓓ 27

Name ....

ESSENTIAL QUESTION ?
How are patterns used in mathematics?

An expression is a combination of numbers, variables, and at least one operation.

When an expression has more than one operation, the order of operations tells us which one to perform first, so that everyone gets the same answer. Parentheses () are symbols which indicate how to group operations.

 Math in My World Watch ▶ | Tutor 💬

Example 1

The table shows how much movie tickets cost. How much will it cost to buy 3 adult tickets and 5 child tickets?

Ticket	Cost
Adult	$8
Child	$5

cost = 3 adult tickets + 5 child tickets

$c = 3 \times \$8 + 5 \times \$5$

$c = \$ \boxed{} + \$ \boxed{}$ First, multiply 3 by $8 and 5 by $5.

$c = \quad \$ \boxed{}$ Add the products to find the total cost.

So, the total cost is $ _____ .

Key Concept   Order of Operations

1. Perform operations in parentheses.
2. Multiply and divide in order from left to right.
3. Add and subtract in order from left to right.

Remember, an expression is a combination of numbers, variables, and at least one operation.

Example 2

Find the value of the expression 3 × (4 + 6).

$3 \times (4 + 6)$

3 × ☐   Perform the operation in the parentheses first.

☐   Multiply.

So, 3 × (4 + 6) = _____ .

Example 3

Find the value of the expression (7 − 3) ÷ (2 + 2).

$(7 - 3) \div (2 + 2)$

☐   ☐   Perform the operations in the parentheses first.

☐   Divide.

So, (7 − 3) ÷ (2 + 2) = _____ .

Guided Practice

Find the value of each expression.

1. 12 − 1 × 3 = _____

2. 15 − 4 × 2 = _____

3. (15 − 4) × 2 = _____

Talk MATH

Explain why Exercises 2 and 3 have different answers even though the numbers are the same.

Name ....

Independent Practice

Find the value of each expression.

4. $8 + 5 \times 2 =$ _____

5. $10 - 1 \times 5 =$ _____

6. $4 + 6 \div 2 =$ _____

7. $9 \times 2 - 6 =$ _____

8. $(16 + 2) \div 3 =$ _____

9. $6 \times (6 - 2) =$ _____

10. $(12 - 4) \div 4 =$ _____

11. $12 - (4 \div 4) =$ _____

12. $(3 + 6) \div (3 \times 1) =$ _____

13. $3 + (6 \div 3) \times 1 =$ _____

Algebra Use the order of operations to find the unknown
in each equation.

14. $5 \times 4 - \blacksquare = 13$

The unknown is _____.

15. $\blacksquare \times (8 + 6) = 42$

The unknown is _____.

16. $(2 + 1) \times (9 - \blacksquare) = 12$

The unknown is _____.

17. $(10 \div 2) + (\blacksquare + 3) = 40$

The unknown is _____.

Problem Solving

18. Each bag of dried apples has 5 servings. Each bag of dried apricots has 3 servings. How many servings of dried fruit are in 6 bags of dried apples and 2 bags of dried apricots?

_____

19. Each book costs $4. How much does it cost to buy 3 books and one magazine that costs $5?

_____

20. **Processes &Practices** **2** **Use Number Sense** A sandwich costs $6 and a drink costs $3. How much does it cost to buy 4 sandwiches and 4 drinks?

_____

Brain Builders

21. **Processes &Practices** **1** **Keep Trying** Use each of the numbers 1, 2, 3, and 4 exactly once in the equation below to make the equation true.

$$(\boxed{} \times \boxed{}) + (\boxed{} \div \boxed{}) = 10$$

22. **Processes &Practices** **1** **Make a Plan** Find possible unknown values to make the equation true. $(\boxed{} \times \boxed{}) + (\boxed{} \div 2) = 15$

Explain why there is more than one possible answer.

_____

23. **Building on the Essential Question** Write a problem that shows why knowing the order of operations is important.

_____

Name ....

MY Homework

Homework Helper

Need help? connectED.mcgraw-hill.com

Find the value of each expression.

$8 \times 7 - (9 \div 3) = ?$

$8 \times 7 - (9 \div 3)$

$8 \times 7 - \quad 3$ Perform the operations in parentheses first.

$56 - \quad 3$ Multiply.

53 Subtract.

So, $8 \times 7 - (9 \div 3) = 53$.

$24 - 2 + 6 \times 3 = ?$

$24 - 2 + 6 \times 3$

$24 - 2 + \quad 18$ Multiply.

$22 + \quad 18$ Subtract.

40 Add.

So, $24 - 2 + 6 \times 3 = 40$.

Practice

Find the value of each expression.

1. $5 + 9 \div 3 =$ _____

2. $46 - (6 \times 5) =$ _____

Find the value of each expression.

3. $(3 + 1) + 27 \div 9 =$ _____

4. $5 \times 5 - 8 =$ _____

5. $(4 + 20) \div 2 + 6 =$ _____

6. $2 \times 9 + 14 \div 2 =$ _____

Problem Solving

7. **Processes & Practices** **Model Math** Tami buys two books that cost $14 each. She pays an additional $2 in tax. How much did Tami pay all together?

_____

Brain Builders

8. Claudio had 34 toy cars. He lost two at the park. Then he divided the rest of the cars evenly among himself and 3 cousins. Write an expression to represent the situation. How many cars did each child get?

_____

9. Last week, Jean did two sit-ups on Monday and three sit-ups on Wednesday. This week, Jean did three times as many sit-ups as last week. Jean used the expression $2 + 3 \times 3$ to find the number of sit-ups she did this week. Explain and correct her error.

_____

10. **Test Practice** Which expression has a value of 20?

Ⓐ $2 \times 5 + 5$

Ⓒ $3 \times 7 - 1$

Ⓑ $3 \times (5 + 5)$

Ⓓ $40 \div 5 - 3$

Check My Progress

Vocabulary Check

Use the words in the word bank to complete each sentence.

equation	input	operation
output	unknown	

1. In the equation $4 + x = 7$, the variable x is a(n) _____.

2. In the table to the right, the letter m

   represents the _____. The letter n

   represents the _____.

3. A(n) _____ is a sentence
   that contains an equals sign (=), showing that two
   expressions are equal.

4. Addition is an example of a(n) _____.

$m + 5 = n$	
m	n
2	7
3	8

Concept Check

**Write an equation that describes the pattern. Then use
the equation to find the next three output numbers.**

5.

Input (a)	4	5	6	7	8	9
Output (b)	9	10	11			

Equation: _____

6.

Input (c)	6	8	10	12	14	16
Output (d)	12	16	20			

Equation: _____

Find the value of each expression.

7. $(7 + 5) \div 3 =$ _____

8. $11 - 2 \times 5 =$ _____

Problem Solving

9. The amount in dollars c a bus company charges to take s students on a field trip are shown at the right. Write an equation to describe the pattern. Then complete the table to show how much it would cost for 40 and 50 students to go on a field trip.

Students	Cost ($)
10	60
20	70
30	80
40	
50	

Brain Builders

10. A local sports team sells 6 tickets for $42, 8 tickets for $56, and 10 tickets for $70. Write a rule and equation to find the cost of 20 tickets.

11. Each peanut butter snack costs $2. Each chocolate snack costs $1 more than each peanut butter snack. How much does it cost to buy 6 peanut butter snacks and 8 chocolate snacks? Write an equation.

12. Test Practice What is the value of m in the equation to the right if $n = 6$?

$$9 \times n = m$$

Ⓐ 15

Ⓒ 54

Ⓑ 27

Ⓓ 81

Lesson 8
Hands On
Equations with Two Operations

ESSENTIAL QUESTION
How are patterns used
in mathematics?

Sometimes an equation has more than one operation.

Build It

Model the equation $(n \times 3) + 5 = y$.

 Set up an equation machine.
Use paper plates to represent the variables, rubber bands for parentheses, paper clips for the equals sign, and index cards to show the numbers and operations.

 Input counters to find y.
Suppose n equals 1. Place 1 counter on the plate labeled n. Move the counters through the machine, following the operations given. Use the order of operations.

$$1 \times 3 = 3 \qquad 3 + 5 = 8 \qquad y = 8$$

 Record the equation.
Fold a piece of paper in half and label it as shown.

When n is...   y equals...
    1   8

So, when $n = 1$, $y = $ _____.

Input more values for the equation (n × 3) + 5 = y. Find new input and output values.

Suppose *n* equals 2. Find the value of *y*.

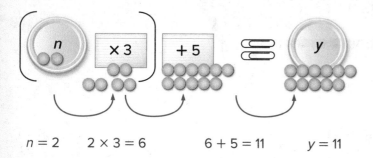

$n = 2$ $2 \times 3 = 6$ $6 + 5 = 11$ $y = 11$

So, when $n = 2$, $y =$ _____ .

Repeat the process for input values of 3, 4, and 5.

Record your input (*n*) and output (*y*) values.

When *n* is...	*y* equals...
1	8
2	
3	
4	
5	

Processes
1. **&Practices** ▶ 2 **Use Algebra** Refer to the equation (*n* × 3) + 5 = *y*. What is the value of *y* when *n* equals 6? 7?

_____

2. Given the equation (*n* + 7) × 3 = *y*, how would you find the value of *y* if *n* equals 3?

_____

_____

Name ....

Use each equation to find each unknown. Use models if needed.

3. $(t + 8) \times 2 = s$

When $t = 4$, $s =$ _____ .

4. $(m \times 6) + 4 = d$

When $m = 3$, $d =$ _____ .

5. $8 + (z \times 2) = w$

When $z = 5$, $w =$ _____ .

6. $(a \div 6) + 5 = b$

When $a = 18$, $b =$ _____ .

7. $12 - (e \times 4) = f$

When $e = 2$, $f =$ _____ .

8. $(r + 8) \times 6 = s$

When $r = 3$, $s =$ _____ .

9. $(g - 4) \times 7 = h$

When $g = 12$, $h =$ _____ .

10. $64 \div (p + 4) = q$

When $p = 4$, $q =$ _____ .

Apply It

11. Set up an equation machine to show $(x + 4) \div 3 = y$.
Find the values of y when $x = 8$, $x = 11$, and $x = 20$.

When $x = 8$, $y =$ _____ .

When $x = 11$, $y =$ _____ .

When $x = 20$, $y =$ _____ .

Processes & Practices **3** **Find the Error** Robert is finding the output values for the equation $a + 7 \times 3 = b$. He wrote a few statements about the input and output values.

12.

> When $a = 3$, $b = 30$.
>
> When $a = 4$, $b = 33$.
>
> When $a = 5$, $b = 36$.

What is Robert's mistake?

_____

_____

Use the equation correctly to find the output values.

> When $a = 3$, $b =$ _____ .
>
> When $a = 4$, $b =$ _____ .
>
> When $a = 5$, $b =$ _____ .

Rewrite the equation so that Robert's values are correct.

_____

Write About It

13. How do parentheses affect the value of expressions?

_____

_____

_____

Name _____

MY Homework

Homework Helper

Need help? connectED.mcgraw-hill.com

Use the equation $(t \times 3) + 5 = w$ to find w when $t = 4$.

$(t \times 3) + 5 = w$

$(4 \times 3) + 5 = w \qquad t = 4$

$12 + 5 = w \qquad$ Perform the operation inside parentheses.

$17 = w \qquad$ Add.

When $t = 4$, $w = 17$.

The counters at the right
model this equation.

Practice

**Use each equation to find each unknown. Draw models
if needed.**

1. $(z + 3) \times 2 = y$

When $z = 2$, $y =$ _____.

2. $4 + (g \times 3) = m$

When $g = 3$, $m =$ _____.

Use each equation to find each unknown. Draw models if needed.

3. $2 + (n \times 7) = p$

When $n = 1$, $p =$ _____ .

4. $(r \times 2) + 6 = v$

When $r = 4$, $v =$ _____ .

5. $6 + (a \times 3) = b$

When $a = 5$, $b =$ _____ .

6. $(j \div 4) + 8 = k$

When $j = 16$, $k =$ _____ .

Problem Solving

7. Set up an equation machine to show $(x + 2) \times 5 = y$. Find the values of y when $x = 5$, $x = 8$, and $x = 12$.

When $x = 5$, $y =$ _____ .

When $x = 8$, $y =$ _____ .

When $x = 12$, $y =$ _____ .

Processes
8. &Practices **2** **Understand Symbols** Bryan made up a game. Each team starts with seven points. Each time a team answers a question correctly, they earn five points. The equation used to find the total number of points is $7 + (5 \times q) = t$. Find the total number of points (t) when a team answers six questions (q) correctly.

_____

Lesson 9
Equations with Multiple Operations

ESSENTIAL QUESTION
How are patterns used
in mathematics?

You have used tables to show equations with one operation. A table can also help you show equations with two operations.

 Math in My World Watch Tutor

Example 1

Sam earns $7 each time he rakes his neighbor's lawn. He also earns $5 each week for doing chores at home. Sam wants to find what he will earn in a week if he does chores and rakes the lawn 1, 2, and 3 times.

 Write an equation.

dollars earned from raking | dollars earned from chores

$$(7 \times t) + 5 = m$$

times he rakes the lawn | money

 Complete the table.

So, during one week, if Sam rakes the lawn

once, he will earn $ _____ . If he rakes

the lawn twice, he will earn $ _____ .

If he rakes the lawn three times,

he will earn $ _____ .

Sam's Money		
Input (*t*)	(7 × *t*) + 5 = *m*	Output (*m*)
1	(7 × 1) + 5 = 12	12
2	(7 × 2) + 5 = 19	
3	(7 × 3) + 5 =	

Equations can have multiple operations.

Example 2

Complete the table to find output numbers when $x = 2, 3, 4,$ and 5.

Helpful Hint

Solve what is between the parentheses first.

$2 \times (9 - x) + 3 = y$	
Input (x)	Output (y)
1	19
2	
3	
4	
5	

Find the value of y when $x = 2$.

$2 \times (9 - 2) + 3 = y$

$2 \times \quad 7 \quad + 3 = y$

$14 \quad + 3 = y$

$17 \quad = y$

Repeat the process when $x = 3, 4,$ and 5.

When $x = 3$, $y =$ _____.

When $x = 4$, $y =$ _____.

When $x = 5$, $y =$ _____.

Describe patterns you see in the table.

_____

_____

Explain how tables can help you solve a problem.

Guided Practice

1. Complete the table.

$(5 + x) \times 4 = y$	
Input (x)	Output (y)
1	24
2	
3	
4	

Independent Practice

Complete each table.

2.

$(7 - x) \times 7 = y$	
Input (x)	Output (y)
1	42
2	
3	
4	

3.

$(2 + x) \times 6 = y$	
Input (x)	Output (y)
1	18
2	
3	
4	

4.

$(4 \times x) - 3 = y$	
Input (x)	Output (y)
1	1
2	
3	
4	

5.

$(9 - x) + 2 = y$	
Input (x)	Output (y)
1	10
2	
3	
4	

6.

$(12 \div x) + 5 = y$	
Input (x)	Output (y)
1	17
2	
3	
4	

7.

$(14 - x) \div 2 = y$	
Input (x)	Output (y)
2	6
4	
6	
8	

8.

$(5 \times x) \div 5 + 1 = y$	
Input (x)	Output (y)
1	2
2	
3	
4	

9.

$3 \times (10 - x) + 4 = y$	
Input (x)	Output (y)
1	31
3	
5	
7	

Problem Solving

Processes &Practices

10. **5** **Use Math Tools** It costs $3 to park at the fair. Tickets cost $6 each. How much will it cost a family of 4 to go to the fair? Complete the table to solve.

_____

($6 × x) + $3 = y	
Input (x)	Output (y)
1	$9
2	
3	
4	

Brain Builders

11. Tam walks 2 miles each way to school each day. During gym class, she always runs three times as far as Dante. How many miles will Tam walk and run in one 5-day school week if Dante runs 1 mile each day?

_____

Processes &Practices

12. **3** **Find the Error** Ashley completed the table shown. Find and correct her mistake. Explain.

_____

_____

(10 − x) × 2 = y	
Input (x)	Output (y)
1	11
2	10
3	9

Write a situation that could be represented by the values in the table.

_____

_____

_____

13. **?** **Building on the Essential Question** Describe a real-world situation that could use a table with two operations. Create a table for the situation with 3 pairs of input/output values.

_____

_____

Name

MY Homework

Homework Helper

Need help? connectED.mcgraw-hill.com

Lauren's recipe calls for 2 times as many cups of flour as sugar.
She always adds 1 cup of oatmeal. If she uses 2, 3, or 4 cups of sugar,
how many cups of flour and oatmeal mix will she use?

1 Write an equation.

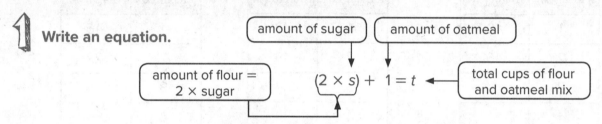

amount of sugar   amount of oatmeal

amount of flour =
2 × sugar

$(2 \times s) + 1 = t$

total cups of flour
and oatmeal mix

2 Make a table.

$(2 \times s) + 1 = t$	
Input (x)	Output (t)
2	5
3	7
4	9

$(2 \times 2) + 1 = 5$

$(2 \times 3) + 1 = 7$

$(2 \times 4) + 1 = 9$

If she uses 2 cups of sugar, she will use 5 cups of flour and oatmeal mix.
If she uses 3 cups of sugar, she will use 7 cups of flour and oatmeal mix.
If she uses 4 cups of sugar, she will use 9 cups of flour and oatmeal mix.

Practice

1. Complete the table.

$(3 \times x) + 2 = y$	
Input (x)	Output (y)
1	5
2	8
3	
4	

Complete each table.

2.

$(12 \div x) + 3 = y$	
Input (x)	Output (y)
1	15
2	9
3	
4	

3.

$(4 + x) \times 6 = y$	
Input (x)	Output (y)
1	30
2	36
3	
4	

4.

$(10 - x) \times 7 = y$	
Input (x)	Output (y)
1	63
2	56
3	
4	

5.

$(5 \times x) + 5 = y$	
Input (x)	Output (y)
1	10
2	15
3	
4	

6.

$(6 + x) \times 2 + 3 = y$	
Input (x)	Output (y)
1	17
2	19
3	
4	

7.

$2 \times (24 \div x) - 2 = y$	
Input (x)	Output (y)
1	46
2	22
3	
4	

Brain Builders

Processes &Practices ① **Make Sense of Problems** Mauricio hits a baseball 4 times as often as Tony each game. He also hits 20 baseballs every Monday at practice. Tony hits 4 balls at the game Saturday. Write and solve an equation to find the number of baseballs Mauricio hit this week.

9. Callie loves flowers. She picks 4 tulips for every daisy she picks. Callie's mom also gave her 6 tulips this week from her garden. How many flowers will Callie have this week if she picks 3 daisies? Explain.

10. **Test Practice** Refer to the equation $(x \times 3) - 2 = y$. If $x = 7$, what is the value of y?

Ⓐ $y = 27$ Ⓑ $y = 23$ Ⓒ $y = 21$ Ⓓ $y = 19$

Write the letter of the definition next to the correct word.

1. **equation** _____

2. **input** _____

3. **nonnumeric pattern** _____

4. **numeric pattern** _____

5. **operation** _____

6. **output** _____

7. **pattern** _____

8. **rule** _____

9. **sequence** _____

10. **term** _____

11. **unknown** _____

A. A statement that describes a relationship between numbers or objects

B. A pattern that uses numbers

C. A sentence that contains an equals sign (=), showing that two expressions are equal

D. An amount that is not known

E. A mathematical process such as addition, subtraction, multiplication, or division

F. A collection of terms that show a pattern

G. A sequence showing a relationship among terms that are not numbers

H. Each number in a sequence

I. The result of an input quantity being changed by a function

J. A sequence of terms that follow a certain order

K. A quantity that is changed by a function to produce an output

Concept Check

12. Extend the pattern. Draw the shape on the line.

         _____

Identify, describe, and extend each pattern.

13. 4, 20, 100, 500, _____

The pattern is _____

_____

14. 44, 22, 20, 10, 8, _____

The pattern is _____

_____

Extend each pattern by four terms. Write an observation about the pattern.

15. Rule: add 6

Pattern: 3, _____, _____, _____, _____

Observation:

_____

_____

16. Rule: multiply by 2

Pattern: 4, _____, _____, _____, _____

Observation:

_____

_____

Write an equation that describes the pattern. Then use the equation to find the next two output numbers.

17.

Input (j)	25	35	45	55	65
Output (k)	21	31	41		

Equation: _____

18.

Input (g)	1	2	3	4	5
Output (h)	3	6	9		

Equation: _____

Find the value of each expression.

19. $7 + 3 \times 6 =$ _____

20. $(6 - 4) \times 9 =$ _____

Find the unknown.

21. $(f + 5) \times 3 = g$

When $f = 4$, $g =$ _____ .

22. $(x \times 4) + 7 = y$

When $x = 8$, $y =$ _____ .

23. Kenneth displays his picture frames in the pattern shown below. Every other picture is of his friends and the rest are of his family. If the first picture is of his family, what picture will be in the third square frame?

24. The admission for an art museum costs $5 per person. Complete the table using the relationship between the number of people and the cost of admission to the museum.

People	Cost
2	
	$15
4	
5	
	$45

25. Mrs. Brown's class can earn 5 minutes of extra recess for each marble she puts in a jar. The class has 15 minutes of recess each morning. Make an input/output table to find how many minutes of recess they will get if they earn 3 marbles. Explain how the order of operations is used to find the amount of recess time.

26. Test Practice Find the value of the expression $(5 + 2) \times 7$.

Ⓐ 14   Ⓒ 21

Ⓑ 19   Ⓓ 49

Reflect

Use what you learned about patterns to
complete the graphic organizer.

ESSENTIAL QUESTION

How are patterns
used in mathematics?

Vocabulary	Nonnumeric Patterns	Numeric Patterns

Reflect on the ESSENTIAL QUESTION Write your answer below.

_____

_____

Performance Task

Farmer's Market

Hattie is selling some of her produce and other foods at the Farmer's Market.

Show all your work to receive full credit.

Part A

Hattie makes a display of boxes of her homemade whole grain macaroni. The first three rows of her display are shown. Extend the pattern by drawing two more rows of boxes.

Row 3

Row 2

Row 1

Part B

Write an equation that is a rule for finding *b*, the number of boxes in row *r* of the display. Explain how you figured out the equation.

Part C

Use your equation to find the number of boxes in Row 7. Show your work.

Part D

How many rows will there be in the display when Hattie is finished making the macaroni display? Explain how you know.

Part E

Hattie made another display with baskets of pears she grew. The number of pears in the baskets form a pattern. The first three baskets are shown. Complete the table. Then state a rule for the pattern.

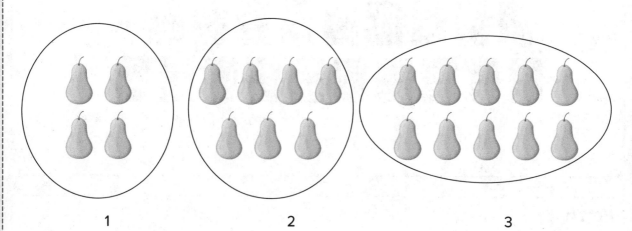

1   2   3

Basket Number	1	2	3	4	5	6
Number of Pears						

Rule: _____

Glossary/Glosario

← Go online for the eGlossary.

Go to the eGlossary to find out more about these words in the following 13 languages:

Arabic · Bengali · Brazilian Portuguese · Cantonese · English · Haitian Creole
Hmong · Korean · Russian · Spanish · Tagalog · Urdu · Vietnamese

English	Spanish/Español
acute angle An *angle* with a measure greater than 0° and less than 90°.	**ángulo agudo** *Ángulo* que mide más de 0° y menos de 90°.

English	Spanish/Español
acute triangle A *triangle* with all three *angles* less than 90°.	**triángulo acutángulo** *Triángulo* cuyos tres *ángulos* miden menos de 90°.

add (adding, addition)   An *operation* on two or more *addends* that results in a *sum*.

$$9 + 3 = 12$$

suma (sumar)   *Operación* de dos o más *sumandos* que da como resultado una *suma*.

$$9 + 3 = 12$$

addend   Any numbers being *added* together.

sumando   Cualquier número que se *suma* a otro.

algebra   A branch of mathematics that uses symbols, usually letters, to explore relationships between quantities.

álgebra   Rama de las matemáticas en la que se usan símbolos, generalmente letras, para explorar relaciones entre cantidades.

Aa

angle A figure that is formed by two *rays* with the same *endpoint*.

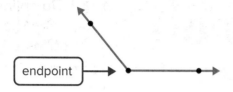

area The number of *square units* needed to cover the inside of a region or plane figure without any overlap.

area = 6 square units

Associative Property of Addition The property that states that the grouping of the *addends* does not change the *sum*.

$$(4 + 5) + 2 = 4 + (5 + 2)$$

Associative Property of Multiplication The property that states that the grouping of the *factors* does not change the *product*.

$$3 \times (6 \times 2) = (3 \times 6) \times 2$$

ángulo Figura formada por dos *semirrectas* con el mismo *extremo*.

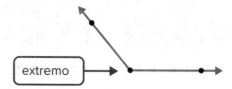

área Cantidad de *unidades cuadradas* necesarias para cubrir el interior de una región o figura plana sin superposiciones.

área = 6 unidades cuadradas

propiedad asociativa de la suma Propiedad que establece que la agrupación de los *sumandos* no altera la *suma*.

$$(4 + 5) + 2 = 4 + (5 + 2)$$

propiedad asociativa de la multiplicación Propiedad que establece que la agrupación de los *factores* no altera el *producto*.

$$3 \times (6 \times 2) = (3 \times 6) \times 2$$

Bb

bar graph A graph that compares *data* by using bars of different *lengths* or heights to show the values.

gráfica de barras Gráfica en la que se comparan los *datos* usando barras de distintas *longitudes* o alturas para mostrar los valores.

capacity The amount of liquid a container can hold.

centimeter (cm) A *metric* unit for measuring *length*.

100 centimeters = 1 meter

circle A closed figure in which all points are the same distance from a fixed point, called the center.

Commutative Property of Addition The property that states that the order in which two numbers are *added* does not change the *sum*.

12 + 15 = 15 + 12

Commutative Property of Multiplication The property that states that the order in which two numbers are *multiplied* does not change the *product*.

7 × 2 = 2 × 7

compatible numbers Numbers in a problem or related numbers that are easy to work with mentally.

720 and 90 are compatible numbers for *division* because 72 ÷ 9 = 8.

composite number A whole number that has more than two factors.

12 has the factors 1, 2, 3, 4, 6, and 12.

capacidad Cantidad que puede contener un recipiente, medida en unidades de volumen.

centímetro (cm) Unidad métrica de *longitud*.

100 centímetros = 1 metro

círculo Figura cerrada en la cual todos los puntos equidistan de un punto fijo llamado centro.

propiedad conmutativa de la suma Propiedad que establece que el orden en el que se *suman* dos o más números no altera la *suma*.

12 + 15 = 15 + 12

propiedad conmutativa de la multiplicación Propiedad que establece que el orden en el que se *multiplican* dos o más números no altera el *producto*.

7 × 2 = 2 × 7

números compatibles Números en un problema o números relacionados con los cuales es fácil trabajar mentalmente.

720 ÷ 90 es una división que usa números compatibles porque 72 ÷ 9 = 8.

número compuesto Número natural con más de dos factores.

12 tiene los factores 1, 2, 3, 4, 6 y 12.

Cc

congruent figures   Two figures having the same size and the same shape.

convert   To change one unit to another.

cup (c)   A *customary* unit of *capacity* equal to 8 fluid ounces.

customary system   The measurement system most often used in the United States. Units include *foot*, *pound*, and *quart*.

figuras congruentes   Dos figuras con la misma forma y el mismo tamaño.

convertir   Cambiar de una unidad a otra.

taza (tz)   Unidad *usual* de *capacidad* que equivale a 8 onzas líquidas.

sistema usual   Conjunto de unidades de medida de uso más frecuente en Estados Unidos. Incluye unidades como *el pie*, *la libra* y *el cuarto*.

Dd

data   Numbers or symbols, sometimes collected from a *survey* or experiment, to show information. Datum is singular; data is plural.

decimal   A number with one or more digits to the right of the decimal point, such as 8.37 or 0.05.

decimal equivalents   Decimals that represent the same number.

$$0.3 \text{ and } 0.30$$

decimal point   A period separating the ones and the *tenths* in a decimal number.

$$0.8 \text{ or } \$3.77$$

decompose   To break a number into different parts.

degree (°)   **a.** A unit for measuring angles. **b.** A unit of measure used to describe temperature.

datos   Números o símbolos que muestran información, algunas veces recopilados a partir de una *encuesta* o un experimento.

decimal   Número con uno o más dígitos a la derecha del *punto decimal,* como 8.37 o 0.05.

decimales equivalentes   Decimales que representan el mismo número.

$$0.3 \text{ y } 0.30$$

punto decimal   Punto que separa las unidades de las *décimas* en un número decimal.

$$0.8 \text{ o } \$3.77$$

descomponer   Separar un número en diferentes partes.

grado (°)   **a.** Unidad que se usa para medir ángulos. **b.** Unidad de medida que se usa para describir la temperatura.

denominator The bottom number in a *fraction*.

In $\frac{5}{6}$, 6 is the denominator.

digit A symbol used to write numbers. The ten digits are 0, 1, 2, 3, 4, 5, 6, 7, 8, and 9.

Distributive Property To *multiply* a *sum* by a number, multiply each *addend* by the number and *add* the *products*.

$$4 \times (1 + 3) = (4 \times 1) + (4 \times 3)$$

dividend A number that is being *divided*.

$3\overline{)19}$ 19 is the dividend.

division (divide) An *operation* on two numbers in which the first number is split into the same number of equal groups as the second number.

divisor The number by which the *dividend* is being *divided*.

$3\overline{)19}$ 3 is the divisor.

denominador El número de abajo en una *fracción*.

En $\frac{5}{6}$, 6 es el denominador.

dígito Símbolo que se usa para escribir los números. Los diez dígitos son 0, 1, 2, 3, 4, 5, 6, 7, 8 y 9.

propiedad distributiva Para *multiplicar* una *suma* por un número, multiplica cada *sumando* por ese número y luego *suma* los *productos*.

$$4 \times (1 + 3) = (4 \times 1) + (4 \times 3)$$

dividendo Número que se *divide*.

$3\overline{)19}$ 19 es el dividendo.

división (dividir) *Operación* entre dos números en la que el primer número se separa en tantos grupos iguales como indica el segundo número.

divisor Número entre el cual se *divide* el *dividendo*.

$3\overline{)19}$ 3 es el divisor.

Ee

elapsed time The amount of time that has passed from beginning to end.

endpoint The point at either end of a *line segment* or the point at the beginning of a *ray*.

endpoints

tiempo transcurrido Cantidad de tiempo que ha pasado entre el principio y el fin de algo.

extremo Punto que se encuentra en cualquiera de los dos lados en que termina un *segmento de recta* o al principio de una *semirrecta*.

extremos

equation   A sentence that contains an equals sign (=), showing that two *expressions* are equal.

ecuación   Espresión matemática que contiene el signo igual, (=), que indica que las dos *expresiones* son iguales.

equilateral triangle   A *triangle* with three *congruent* sides.

triángulo equilátero   *Triángulo* con tres lados *congruentes*.

equivalent fractions   *Fractions* that represent the same number.

$$\frac{3}{4} = \frac{6}{8}$$

fracciones equivalentes   *Fracciones* que representan el mismo número.

$$\frac{3}{4} = \frac{6}{8}$$

estimate   A number close to an exact value. An estimate indicates *about* how much.

$$47 + 22 \text{ is about}$$
$$50 + 20 \text{ or } 70.$$

estimación   Número cercano a un valor exacto. Una estimación indica una cantidad approximada.

$$47 + 22 \text{ es aproximadamente}$$
$$50 + 20 \text{ o } 70.$$

expanded form/expanded notation The representation of a number as a *sum* that shows the value of each digit.

536 is written as 500 + 30 + 6.

forma desarrollada/notación desarrollada   Representación de un número como la *suma* del valor de cada dígito.

536 puede escribirse como 500 + 30 + 6.

expression   A combination of numbers, *variables,* and at least one *operation.*

expresión   Combinación de números, *variables* y por lo menos una *operación.*

fact family   A group of related facts using the same numbers.

5 + 3 = 8	5 × 3 = 15
3 + 5 = 8	3 × 5 = 15
8 − 3 = 5	15 ÷ 3 = 5
8 − 5 = 3	15 ÷ 5 = 3

familia de operaciones   Grupo de operaciones relacionadas que usan los mismos números.

5 + 3 = 8	5 × 3 = 15
3 + 5 = 8	3 × 5 = 15
8 − 3 = 5	15 ÷ 3 = 5
8 − 5 = 3	15 ÷ 5 = 3

factor   A number that *divides* a whole number evenly. Also a number that is *multiplied* by another number.

factor   Número entre el que se *divide* otro número natural sin dejar residuo. También es cada uno de los números en una multiplicación.

factor pairs   The two factors that are multiplied to find a product.

pares de factores   Los dos factores que se multiplican para hallar un producto.

fluid ounce (fl oz)   A *customary* unit of *capacity*.

onza líquida (oz líq)   Unidad *usual* de *capacidad*.

foot (ft)   A *customary* unit for measuring *length*. Plural is feet.

1 foot = 12 inches

pie (pie)   Unidad *usual* de *longitud*.

1 pie = 12 pulgadas

formula   An *equation* that shows the relationship between two or more quantities.

fórmula   *Ecuación* que muestra la relación entre dos o más cantidades.

fraction   A number that represents part of a whole or part of a set.

$$\frac{1}{2}, \frac{1}{3}, \frac{1}{4}, \frac{3}{4}$$

fracción   Número que representa una parte de un todo o una parte de un conjunto.

$$\frac{1}{2}, \frac{1}{3}, \frac{1}{4}, \frac{3}{4}$$

frequency table   A table for organizing a set of *data* that shows the number of times each result has occurred.

tabla de frecuencias   Tabla para organizar un conjunto de *datos* que muestra el número de veces que se ha obtenido cada resultado.

Gg

gallon (gal)   A *customary* unit for measuring *capacity* for liquids.

1 gallon = 4 quarts

galón (gal)   Unidad *usual* de *capacidad* de líquidos.

1 galón = 4 cuartos

gram (g)   A *metric* unit for measuring *mass*.

gramo (g)   Unidad *métrica* de *masa*.

Gg

Greatest Common Factor (GCF)
The greatest of the common *factors* of two
or more numbers.

> The greatest common factor of
> 12, 18, and 30 is 6.

máximo común divisor (M.C.D.)
El mayor de los *factores* comunes de dos
o más números.

> El máximo común divisor de
> 12, 18 y 30 es 6.

Hh

hexagon   A *polygon* with six sides and six
angles.

hexágono   *Polígono* con seis lados
y seis *ángulos.*

hundredth   A *place-value* position. One
of one hundred equal parts.

> In the number 0.05, 5 is in
> the hundredths place.

centésima   *Valor posicional.*
Una de cien partes iguales.

> En el número 4.57, 7 está en
> el lugar de las centésimas.

Ii

Identity Property of Addition   For any
number, zero plus that number is the
number.

> $3 + 0 = 3$ or $0 + 3 = 3$

propiedad de identidad de la suma
Para todo número, cero más ese número
da como resultado ese mismo número.

> $3 + 0 = 3$ o $0 + 3 = 3$

Identity Property of Multiplication
If you *multiply* a number by 1, the
product is the same as the given
number.

> $8 \times 1 = 8 = 1 \times 8$

**propiedad de identidad de la
multiplicación**   Si *multiplicas* un número
por 1, el *producto* es igual al número dado.

> $8 \times 1 = 8 = 1 \times 8$

improper fraction   A *fraction* with a
numerator that is greater than or equal to
the *denominator.*

> $\frac{17}{3}$ or $\frac{5}{5}$

fracción impropia   *Fracción* con un
numerador igual al *denominador* o mayor
que él.

> $\frac{17}{3}$ o $\frac{5}{5}$

input A quantity that is changed to produce an output.

intersecting lines *Lines* that meet or cross at a point.

is equal to (=) Having the same value. The (=) sign is used to show two numbers or *expressions* are equal.

is greater than (>) An inequality relationship showing that the number on the left of the symbol is greater than the number on the right.

5 > 3   5 is greater than 3.

is less than (<) An inequality relationship showing that the number on the left side of the symbol is less than the number on the right side.

4 < 7   4 is less than 7.

isosceles triangle A *triangle* with at least 2 sides of the same *length*.

entrada Cantidad que se modifica y produce un valor de salida.

rectas secantes *Rectas* que se intersecan o se cruzan en un punto común.

es igual a (=) Que tiene el mismo valor. Con el signo = se muestra que dos números o *expresiones* son iguales.

es mayor que (>) Relación de desigualdad que muestra que el número a la izquierda del signo es más grande que el número a la derecha.

5 > 3   5 es mayor que 3.

es menor que (<) Relación de desigualdad que muestra que el número a la izquierda del signo es más pequeño que el número a la derecha.

4 < 7   4 es menor que 7.

triángulo isósceles *Triángulo* que tiene por lo menos 2 lados del mismo *largo*.

Kk

kilogram (kg) A *metric* unit for measuring *mass*.

kilometer (km) A *metric* unit for measuring *length*.

kilogramo (kg) Unidad *métrica* de *masa*.

kilómetro (km) Unidad *métrica* de *longitud*.

Ll

length The measurement of a *line* between two points.

longitud Medida de la distancia entre dos puntos.

like fractions *Fractions* that have the same *denominator*.

$$\frac{1}{5} \text{ and } \frac{2}{5}$$

fracciones semejantes *Fracciones* que tienen el mismo *denominador*.

$$\frac{1}{5} \text{ y } \frac{2}{5}$$

line A straight set of points that extend in opposite directions without ending.

recta Conjunto de puntos alineados que se extiende sin fin en direcciones opuestas.

line of symmetry A *line* on which a figure can be folded so that its two halves match exactly.

eje de simetría *Recta* sobre la cual se puede doblar una figura de manera que sus mitades coincidan exactamente.

line plot A graph that uses columns of Xs above a *number line* to show frequency of *data*.

diagrama lineal Gráfica que tiene columnas de X sobre una *recta numérica* para representar la frecuencia de los *datos*.

line segment   A part of a *line* between two *endpoints*. The *length* of the line segment can be measured.

A   B

line symmetry   A figure has *line symmetry* if it can be folded so that the two parts of the figure match, or are *congruent*.

liter (L)   A *metric* unit for measuring *volume* or *capacity*.

1 liter = 1,000 milliliters

segmento de recta   Parte de una *recta* entre dos *extremos*. La *longitud* de un segmento de recta se puede medir.

A   B

simetría axial   Una figura tiene *simetría axial* si puede doblarse de modo que las dos partes de la figura coincidan de manera exacta.

litro (L)   Unidad *métrica* de *volumen* o *capacidad*.

1 litro = 1,000 mililitros

mass   The amount of matter in an object. Two examples of units of measure would be gram and kilogram.

meter (m)   A *metric* unit for measuring *length*.

metric system (SI)   The decimal system of measurement. Includes units such as *meter*, *gram*, and *liter*.

mile (mi)   A *customary* unit of measure for *length*.

1 mile = 5,280 feet

milliliter (mL)   A *metric* unit for measuring *capacity*.

1,000 milliliters = 1 liter

millimeter (mm)   A *metric* unit for measuring *length*.

1,000 millimeters = 1 meter

masa   Cantidad de materia en un cuerpo. El gramo y el kilogramo son dos ejemplos de unidades que se usan para medir la masa.

metro (m)   Unidad *métrica* de *longitud*.

sistema métrico (SI)   Sistema decimal de medidas que se basa en potencias de 10 y que incluye unidades como *el metro*, *el gramo* y *el litro*.

milla (mi)   Unidad *usual* de *longitud*.

1 milla = 5,280 pies

mililitro (mL)   Unidad *métrica* de *capacidad*.

1,000 mililitros = 1 litro

milímetro (mm)   Unidad *métrica* de *longitud*.

1,000 milímetros = 1 metro

Mm

minuend   The first number in a *subtraction* sentence from which a second number is to be subtracted.

mixed number   A number that has a *whole number* part and a *fraction* part.

$$6\frac{3}{4}$$

multiple   A multiple of a number is the *product* of that number and any whole number.

15 is a multiple of 5
because $3 \times 5 = 15$.

multiply (multiplication)   An *operation* on two numbers to find their *product*. It can be thought of as repeated *addition*.

minuendo   El primer número en un enunciado de *resta* del cual se restará un segundo número

$$8 \quad - \quad 3 \quad = \quad 5$$

minuendo   sustraendo   diferencia

número mixto   Número formado por un *número natural* y una parte *fraccionaria*.

$$6\frac{3}{4}$$

múltiplo   Un múltiplo de un número es el *producto* de ese número y cualquier otro número natural.

15 es múltiplo de 5
porque $3 \times 5 = 15$.

multiplicar (multiplicación)   *Operación* entre dos números para hallar su *producto*. También se puede interpretar como una *suma* repetida.

Nn

nonnumeric pattern   Patterns that do not use numbers.

number line   A *line* with numbers on it in order at regular intervals.

numerator   The number above the bar in a *fraction*; the part of the fraction that tells how many of the equal parts are being used.

numeric pattern   Patterns that use numbers.

patrón no numérico   Patrón que no usa números.

recta numérica   *Recta* con números ordenados a intervalos regulares.

numerador   El número que está encima de la barra de *fracción*; la parte de la fracción que indica cuántas de las partes iguales en que se divide el entero se están usando.

patrón numérico   Patrón que usa números.

obtuse angle   An *angle* that measures greater than 90° but less than 180°.

obtuse triangle   A *triangle* with one *obtuse angle.*

octagon   A *polygon* with 8 sides and 8 *angles.*

operation   A mathematical process such as *addition* (+), *subtraction* (−), *multiplication* (×), or *division* (÷).

order of operations   Rules that tell what order to follow when evaluating an *expression:*
(1) Do the *operations* in *parentheses* first.
(2) *Multiply* and *divide* in order from left to right.
(3) *Add* and *subtract* in order from left to right.

ounce (oz)   A *customary* unit to measure *weight* or *capacity.*

output   The result of an input quantity being changed.

ángulo obtuso   *Ángulo* que mide más de 90° pero menos de 180°.

triángulo obtusángulo   *Triángulo* con un *ángulo obtuso.*

octágono   *Polígono* de 8 lados y 8 *ángulos.*

operación   Proceso matemático como la *suma* (+), la *resta* (−), la *multiplicación* (×) o la *división* (÷).

orden de las operaciones   Reglas que te indican qué orden seguir cuando evalúas una *expresión:*
(1) Resuelve primero las *operaciones* dentro de los *paréntesis.*
(2) *Multiplica* o *divide* en orden de izquierda a derecha.
(3) *Suma* o *resta* en orden de izquierda a derecha.

onza (oz)   Unidad *usual* de *peso* o *capacidad.*

salida   Resultado que se obtiene al modificar un valor de entrada.

Pp

parallel lines   *Lines* that are the same distance apart. Parallel lines do not meet.

parallelogram   A *quadrilateral* with four sides in which each pair of opposite sides are *parallel* and equal in *length*.

6 ft
2 ft   2 ft
6 ft

parentheses   The enclosing symbols ( ), which indicate that the terms within are a unit.

partial products   A multiplication method in which the products of each place value are found separately, and then added together.

partial quotients   A dividing method in which the dividend is separated into sections that are easy to divide.

pattern   A sequence of numbers, figures, or symbols that follows a rule or design.

2, 4, 6, 8, 10

pentagon   A *polygon* with five *sides* and five *angles*.

rectas paralelas   *Rectas* separadas por la misma distancia en cualquier punto. Las rectas paralelas no se intersecan.

paralelogramo   *Cuadrilátero* en el que cada par de lados opuestos son *paralelos* y tienen la misma *longitud*.

6 pies
2 pies   2 pies
6 pies

paréntesis   Los signos ( ) con que se encierran los términos, para indicar que cuando están adentro, forman una unidad.

productos parciales   Método de multiplicación por el cual los productos de cada valor posicional se hallan por separado y luego se suman entre sí.

cocientes parciales   Método de división por el cual el dividendo se separa en secciones que son fáciles de dividir.

patrón   Sucesión de números, figuras o símbolos que sigue una regla o un diseño.

2, 4, 6, 8, 10

pentágono   *Polígono* de cinco *lados* y cinco *ángulos*.

percent A ratio that compares a number to 100.

perimeter The distance around a shape or region.

period The name given to each group of three digits on a place-value chart.

perpendicular lines *Lines* that meet or cross each other to form *right angles.*

pint (pt) A *customary* unit for measuring *capacity.*

1 pint = 2 cups

place value The value given to a *digit* by its position in a number.

point An exact location in space that is represented by a dot.

polygon A closed *plane figure* formed using *line segments* that meet only at their *endpoints.*

pound (lb) A *customary* unit to measure *weight* or *mass.*

1 pound = 16 ounces

porcentaje Razón que compara un número con el 100.

perímetro Distancia alrededor de una figura o región.

período Nombre dado a cada grupo de tres dígitos en una tabla de valor posicional.

rectas perpendiculares *Rectas* que se intersecan o cruzan formando *ángulos rectos.*

pinta (pt) Unidad *usual* de *capacidad.*

1 pinta = 2 tazas

valor posicional Valor dado a un *dígito* según su posición en un número.

punto Ubicación exacta en el espacio que se representa con una marca puntual.

polígono *Figura plana* cerrada formada por *segmentos de recta* que solo se unen en sus *extremos.*

libra (lb) Unidad *usual* de *peso* o *masa.*

1 libra = 16 onzas

Pp

prime number   A whole number with exactly two *factors*, 1 and itself.

7, 13, and 19

product   The answer or result of a *multiplication* problem. It also refers to expressing a number as the *product* of its *factors*.

protractor   An instrument used to measure angles.

número primo   Número natural que tiene exactamente dos *factores*: 1 y sí mismo.

7, 13 y 19

producto   Respuesta o resultado de un problema de *multiplicación*. Además, un número puede expresarse como el *producto* de sus *factores*.

transportador   Instrumento con el que se miden los ángulos.

Qq

quadrilateral   A shape that has 4 sides and 4 *angles*.

square, rectangle, and *parallelogram*

cuadrilátero   Figura que tiene 4 lados y 4 *ángulos*.

cuadrado, rectángulo y *paralelogramo*

quart (qt)   A *customary* unit for measuring *capacity*.

1 quart = 4 cups

cuarto (ct)   Unidad *usual* de *capacidad*.

1 cuarto = 4 tazas

quotient   The result of a *division* problem.

cociente   Respuesta o resultado de un problema de *división*.

Rr

ray   A part of a *line* that has one *endpoint* and extends in one direction without ending.

semirrecta   Parte de una *recta* que tiene un *extremo* y que se extiende sin fin en una dirección.

rectangle   A *quadrilateral* with four *right angles*; opposite sides are equal and *parallel*.

rectángulo   *Cuadrilátero* con cuatro *ángulo rectos*; los lados opuestos son iguales y *paralelos*.

regroup   To use place value to exchange equal amounts when renaming a number.

reagrupar   Usar el valor posicional para expresar una cantidad de otra manera.

remainder   The number that is left after one whole number is *divided* by another.

residuo   Número que queda después de *dividir* un número natural entre otro.

repeated subtraction   To subtract the same number over and over until you reach 0.

resta repetida   Procedimiento por el que se resta un número una y otra vez hasta llegar a 0.

rhombus   A *parallelogram* with four *congruent* sides.

rombo   *Paralelogramo* con cuatro lados *congruentes*.

right angle   An *angle* with a measure of 90°.

ángulo recto   *Ángulo* que mide 90°.

right triangle   A *triangle* with one *right angle*.

triángulo rectángulo   *Triángulo* con un *ángulo recto*.

Rr

round   To change the value of a number to one that is easier to work with. To find the nearest value of a number based on a given *place value*.

rule   A statement that describes a relationship between numbers or objects.

redondear   Cambiar el valor de un número a uno con el cual es más fácil trabajar. Hallar el valor más cercano a un número basándose en un *valor posicional* dado.

regla   Enunciado que describe una relación entre números u objetos.

Ss

second   A unit of time.
60 seconds = 1 minute

segundo   Unidad de tiempo.
60 segundos = 1 minuto

sequence   The ordered arrangement of terms that make up a pattern.

secuencia   Disposición ordenada de términos que forman un patrón.

simplest form   A *fraction* in which the *numerator* and the *denominator* have no common *factor* greater than 1.
$\frac{3}{5}$ is the simplest form of $\frac{6}{10}$.

mínima expresión   *Fracción* en la que el *numerador* y el *denominador* no tienen un *factor* común mayor que 1.
$\frac{3}{5}$ es la mínima expresión de $\frac{6}{10}$.

solve   To replace a *variable* with a value that results in a true sentence.

resolver   Despejar una *variable* y verdadera reemplazarla por un valor que haga que la ecuación sea.

square   A *rectangle* with four *congruent* sides.

cuadrado   *Rectángulo* de cuatro lados *congruentes*.

square unit   A unit for measuring *area*.

unidad cuadrada   Unidad para medir el *área*.

standard form/standard notation   The usual way of writing a number that shows only its *digits*, no words.

537   89   1,642

forma estándar/notación estándar
Manera habitual de escribir un número usando solo sus *dígitos*, sin usar palabras.

537   89   1,642

subtract (subtraction) An *operation* on two numbers that tells the *difference*, when some or all are taken away. Subtraction is also used to compare two numbers.

$$14 - 8 = 6$$

subtrahend A number that is *subtracted* from another number.

$$14 - 5 = 9$$

↑

subtrahend

sum The answer to an *addition* problem.

survey A method of collecting *data*.

restar (resta) *Operación* con dos números que indica la *diferencia,* entre ellos. Puede usarse para quitar una cantidad de otra o para comparar dos números.

$$14 - 8 = 6$$

sustraendo Un número que se *resta* de otro número.

$$14 - 5 = 9$$

↑

sustraendo

suma Respuesta o resultado que se obtiene al sumar.

encuesta Método para recopilar *datos.*

tally chart A way to keep track of *data* using *tally marks* to record the number of responses or occurrences.

What is Your Favorite Color?	
Color	Tally
Blue	ⵌⵌ |||
Green	||||

tally mark(s) A mark made to keep track of and display *data* recorded from a *survey.*

tenth One of ten equal parts, or $\frac{1}{10}$.

term Each number in a numeric pattern.

tabla de conteo Manera de llevar la cuenta de los *datos* usando *marcas de conteo* para anotar el número de respuestas o sucesos.

¿Cuál es tu color favorito?	
Color	Conteo
azul	ⵌⵌ |||
verde	||||

marca de conteo Marca que se hace para llevar un registro y representar *datos* recopilados en una *encuesta.*

décima Una de diez partes iguales o $\frac{1}{10}$.

término Cada número en un patrón numérico.

Tt

thousandth(s) One of a thousand equal parts, or $\frac{1}{1000}$. Also refers to a *place value* in a *decimal* number. In the *decimal* 0.789, the 9 is in the thousandths place.

three-dimensional figure A solid figure has three dimensions: *length,* width, and height.

ton (T) A *customary* unit to measure *weight*.

1 ton = 2,000 pounds

trapezoid A *quadrilateral* with exactly one pair of *parallel* sides.

triangle A *polygon* with three sides and three *angles*.

two-dimensional figure A figure that lies entirely within one plane.

milésima Una de mil partes iguales o $\frac{1}{1000}$. También se refiere a un *valor posicional* en un número *decimal*. En el *decimal* 0.789, el 9 está en el lugar de las milésimas.

figura tridimensional Figura sólida que tiene tres dimensiones: *largo,* ancho y alto.

tonelada (T) Unidad *usual* de *peso*.

tonelada = 2,000 libras

trapecio *Cuadrilátero* con exactamente un par de lados *paralelos.*

triángulo *Polígono* con tres lados y tres *ángulos.*

figura bidimensional Figura que puede representarse en un plano.

Uu

unit square A square with a side length of one unit.

unknown The amount that has not been identified.

cuadrado unitario Cuadrado cuyos lados miden una unidad de longitud.

incógnita La cantidad que no ha sido identificada.

Vv

variable   A letter or symbol used to represent an unknown quantity.

Venn diagram   A diagram that uses *circles* to display elements of different sets. Overlapping *circles* show common elements.

Factors of 42   Factors of 56

vertex   The point where two *rays* meet in an *angle.*

variable   Letra o símbolo que se usa para representar una cantidad desconocida.

diagrama de Venn   Diagrama con *círculos* para mostrar elementos de diferentes conjuntos. Los *círculos* sobrepuestos indican elementos comunes.

Factores de 42   Factores de 56

vértice   Punto donde se unen dos *semirrectas* formando un *ángulo.*

Ww

weight   A measurement that tells how heavy an object is.

word form/word notation   The form of a number that uses written words.

peso   Medida que indica cuán pesado o liviano es un cuerpo.

forma verbal/notación verbal   Forma de expresar un número usando palabras escritas.

Yy

yard (yd)   A *customary* unit of *length* equal to 3 feet or 36 inches.

yarda (yd)   Unidad *usual* de *longitud* igual a 3 pies o 36 pulgadas.

Zero Property of Multiplication
The property that states any number *multiplied* by zero is zero.

$$0 \times 5 = 0 \qquad 5 \times 0 = 0$$

propiedad del cero de la multiplicación
Propiedad que establece que cualquier número *multiplicado* por cero es igual a cero.

$$0 \times 5 = 0 \qquad 5 \times 0 = 0$$

Name ....

Work Mat 1: Hundreds Chart

1	2	3	4	5	6	7	8	9	10
11	12	13	14	15	16	17	18	19	20
21	22	23	24	25	26	27	28	29	30
31	32	33	34	35	36	37	38	39	40
41	42	43	44	45	46	47	48	49	50
51	52	53	54	55	56	57	58	59	60
61	62	63	64	65	66	67	68	69	70
71	72	73	74	75	76	77	78	79	80
81	82	83	84	85	86	87	88	89	90
91	92	93	94	95	96	97	98	99	100

Work Mat 2: Thousands, Hundreds, Tens, and Ones Chart

Thousands	Hundreds	Tens	Ones

Name ....

Work Mat 3: Place-Value Chart

Millions			Thousands			Ones		
hundreds	tens	ones	hundreds	tens	ones	hundreds	tens	ones

Work Mat 4: Algebra Mat

=